THAT'S MATHS

Peter Lynch

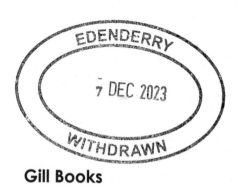

Gill Books

Gill Books
Hume Avenue
Park West
Dublin 12
www.gillbooks.ie

Gill Books is an imprint of M.H. Gill & Co.

978 07171 6955 9

Edited by Jane Rogers
Designed by Síofra Murphy
Printed by TJ International, Cornwall

p. 77: Wikimedia Commons / Edmont; p. 112: Wikimedia Commons / Victor Treushchenko; p. 118: Wikimedia Commons / Aaron Rotenberg; p. 123: (left) Fragment A of the Antikythera Mechanism (Alamy), (right) diagram of the gearing of the AM (Wikimedia Commons / Lead Holder); p. 139: photo from *Mathematical Cuneiform Tablets*, Otto Neugebauer and Abraham Sachs, 1945; p. 146: Alamy; p. 284: Wikimedia Commons / Paris 16; p. 293: Hans Holbein the Younger, *The Ambassadors* (detail of skull seen from right) © The National Gallery, London; p. 305: The Ethel Morrison Van Derlip Fund, proceeds from the 2011 Minneapolis Print and Drawing Fair, the Richard Lewis Hillstrom Fund, gift of funds from Nivin MacMillan, the Winton Jones Endowment Fund for Prints and Drawings, and gift of Herschel V. Jones, gift of Mrs. Ridgely Hunt, and gift of Miss Eileen Bigelow and Mrs. O. H. Ingram, in memory of their mother, Mrs. Alice F. Bigelow, by exchange.

This book is typeset in Century Gothic 10/15 pt.

The paper used in this book comes from the wood pulp of managed forests. For every tree felled, at least one tree is planted, thereby renewing natural resources.

A CIP catalogue record for this book is available from the British Library.

5 4 3 2 1

ACKNOWLEDGEMENTS

My warm thanks go to all my colleagues in the School of Mathematics and Statistics, University College Dublin, for providing ideas and for numerous interesting conversations.

I am also grateful to the staff of *The Irish Times*, especially to the Science Editor, Dick Ahlstrom, for facilitating the publication of mathematical articles in the newspaper.

Finally, I should like to acknowledge the marvellous contributions of thousands of scholars to the online encyclopaedia, Wikipedia – a resource of inestimable value.

PREFACE

This book is a collection of articles covering all major aspects of mathematics. It is written for people who have a keen interest in science and mathematics but who may not have the technical knowledge required to study mathematical texts and journals. The articles are accessible to anyone who has studied mathematics at secondary school.

Mathematics can be enormously interesting and inspiring, but its beauty and utility are often hidden. Many of us did not enjoy mathematics at school and have negative memories of slogging away, trying to solve pointless and abstruse problems. Yet we realise that mathematics is essential for modern society and plays a key role in our economic welfare, health and recreation.

Mathematics can be demanding on the reader because it requires active mental effort. Recognising this, the present book is modular in format. Each article can be read as a self-contained unit. I have resisted the temptation to organise the articles into themes, presenting them instead in roughly the order in which they were written. Each article tells its own story, whether it is a biography of some famous mathematician, a major problem (solved or unsolved), an application of maths to technology or a cultural connection to music or the visual arts.

I have attempted to maintain a reasonably uniform mathematical level throughout the book. You may have forgotten the details of what you learned at school, but what remains should be sufficient to enable you to understand the articles. If you find a particular article abstruse or difficult to understand, just skip to the next one, which will be easier. You can always return later if you wish.

The byline of my blog, thatsmaths.com, is 'Beautiful, Useful and Fun'. I have tried to bring out these three aspects of mathematics in the articles. Beauty can be subjective, but, as you learn more, you cannot fail to be impressed by the majesty and splendour of the intellectual creations of some of the world's most brilliant minds. The usefulness of maths is shown by its many applications to modern technology, and its growing role in medicine, biology and the social sciences. The fun aspect will be seen in the field known as recreational mathematics, aspects of maths that no longer attract active professional research but that still hold fascination.

About half the articles have appeared in *The Irish Times* over the past four years. The remainder are newly written pieces and postings from thatsmaths.com. If you have a general interest in scientific matters and wish to be inspired by the beauty and power of mathematics, this book should serve you well.

CONTENTS

- Introduction   1
- You Can Do Maths   5
- Instant Information   8
- Napier's Nifty Rules   11
- Sproutology   15
- Why Don't Clouds Fall Down?   18
- Packing Oranges and Stacking Cannonballs   21
- Modelling Epidemics   24
- A Falling Slinky   27
- A 'Mersennery' Quest   29
- Shackleton's Spectacular Boat Journey   32
- Where in the World?   35
- Srinivasa Ramanujan   38
- Sharing a Pint   41
- Pons Asinorum   44
- Lost and Found: The Secrets of Archimedes   47
- Subterranean Topology   50
- The Earth's Vast Orb   53
- More Equal than Others   56
- Maths and CAT Scans   58
- Bayes Rules OK   61

▶ Pythagoras goes Global   64

▶ Dozenal Digits: From Dix to Douze   67

▶ How Leopards get their Spots   69

▶ Monster Symmetry and the Forces of Nature   72

▶ Kelvin Wakes   75

▶ Gauss Misses a Trick   78

▶ Prime Secrets Revealed   82

▶ Amazing Normal Numbers   85

▶ Heavy Metal or Blue Jeans?   88

▶ The School of Athens   91

▶ Hailstone Numbers   94

▶ The Remarkable BBP Formula   97

▶ The Atmospheric Railway   101

▶ A Hole through the Earth   104

▶ Sofia Kovalevskaya   107

▶ The Simpler the Better   110

▶ Geometry out of this World   113

▶ Euler's Gem   116

▶ The Watermelon Puzzle   119

▶ The Antikythera Mechanism: The First
  Computer   121

▶ World Population   124

▶ Ireland's Fractal Coast   127

▶ Santa's Fractal Journey   131

▶ Interesting Bores   134

▶ Pythagorean (or Babylonian) Triples   137

▸ Bézout's Theorem 142

▸ French Curves and Bézier Splines 146

▸ Astronomical Perturbations 150

▸ The Predictive Power of Maths 154

▸ Highway Geometry 157

▸ Breaking Weather Records 160

▸ The Faraday of Statistics 163

▸ The Chaos Game 166

▸ Fibonacci Numbers are Good for Business 170

▸ Biscuits, Books, Coins and Cards:
Severe Hangovers 173

▸ Gauss's Great Triangle and the Shape
of Space 176

▸ Degrees of Infinity 181

▸ A Swinging Way to See the Spinning Globe 185

▸ Do You Remember Venn? 188

▸ Mathematics is Coming to Life in a Big Way 191

▸ Temperamental Tuning 194

▸ Cartoon Curves 199

▸ How Big was the Bomb? 202

▸ Algebra in the Golden Age 205

▸ Old Octonions May Rule the World 208

▸ Light Weight 211

▸ Falling Bodies 215

▸ Earth's Shape and Spin Won't Make You Thin 218

▸ The Tangled Tale of Knots 221

▸ Plateau's Problem: Soap Bubbles and Soap Films 224

▸ The Steiner Minimal Tree Problem 229

▸ Who Wants to be a Millionaire? 234

▸ The Klein 4-Group 237

▸ Tracing Our Mathematical Ancestry: The Mathematics Genealogy Project 242

▸ Café Mathematics in Lvov 245

▸ The King of Infinite Space: Euclid and his Elements 248

▸ Golden Moments 253

▸ Mode-S EHS: A Novel Source of Weather Data 256

▸ For Good Communications, Leaky Cables are Best 259

▸ Tap-tap-tap the Cosine Button 262

▸ The Black–Scholes Equation 265

▸ Eccentric Pizza Slices 268

▸ Mercator's Marvellous Map 270

▸ The Remarkable Power of Symmetry 273

▸ Increasingly Abstract Algebra 276

▸ Acoustic Excellence and RT-60 279

▸ The Bridges of Paris 282

▸ Buffon Was No Buffoon 285

▸ James Joseph Sylvester 288

▸ Holbein's Anamorphic Skull 291

▸ The Ubiquitous Cycloid 294
▸ Hamming's Smart Error-correcting Codes 297
▸ Mowing the Lawn in Spirals 300
▸ *Melencolia* I: An Enigma for Half
  a Millennium 303
▸ Mathematics Can Solve Crimes 306
▸ Life's a Drag Crisis 309
▸ The Flight of a Golf Ball 311
▸ Factorial 52: A Stirling Problem 314
▸ Richardson's Fantastic Forecast Factory 317
▸ The Analemmatic Sundial 320
▸ Further Reading 323
▸ Index 325

INTRODUCTION

BEAUTIFUL, USEFUL AND FUN: THAT'S MATHS

Type a word into Google: a billion links come back in a flash. Tap a destination into your satnav: distances, times and highlights of the route appear. Get cash from an ATM, safe from prying eyes. Choose a tune from among thousands squeezed onto a tiny chip. How are these miracles of modern technology possible? What is the common basis underpinning them? The answer is mathematics.

Maths now reaches into every corner of our lives. Our technological world would be impossible without it. Electronic devices like smartphones and iPods, which we use daily, depend on the application of maths, as do computers, communications and the internet. International trade and the financial markets rely critically on secure communications, using encryption methods that spring directly from number theory, once thought to be a field of pure mathematics without 'useful' applications.

We are living longer and healthier lives, partly due to the application of maths to medical imaging, automatic diagnosis and modelling the cardiovascular system. The pharmaceuticals that cure us and control disease are made possible through applied mathematics.

Agricultural production is more efficient thanks to maths; forensic medicine and crime detection depend on it. Control and operation of air transport would be impossible without maths. Sporting records are broken by studying and modelling performance and designing equipment mathematically. Maths is everywhere.

THE LANGUAGE OF NATURE

Galileo is credited with quantifying the study of the physical world, and his philosophy is encapsulated in the oft-quoted aphorism, 'The Book of Nature is written in the language of mathematics.' This development flourished with Isaac Newton, who unified terrestrial and celestial mechanics in a grand theory of universal gravitation, showing that the behaviour of a projectile like a cannonball and the trajectory of the moon are governed by the same dynamics.

Mechanics and astronomy were the first subjects to be 'mathematicised', but over the past century the influence of quantitative methods has spread to many other fields. Statistical analysis now pervades the social sciences. Computers enable us to simulate complex systems and predict their behaviour. Modern weather forecasting is an enormous arithmetical calculation, underpinned by mathematical and physical principles. With the recent untangling of the human genome, mathematical biology is a hot topic.

The mathematics that we learned at school was developed centuries ago, so it is easy to get the idea that maths is static, frozen in the seventeenth century or fossilised since ancient Greece. In fact, the vast bulk of mathematics has emerged in the past hundred years, and the subject continues to blossom. It is a vibrant and dynamic field of study. The future health of our technological society depends on this continuing development.

While a deep understanding of advanced mathematics requires intensive study over a long period, we can appreciate some of the beauty of maths without detailed technical knowledge, just as we can enjoy music without being performers or composers. It is a goal of this book to assist readers in this appreciation. It is hoped that, through this collection of articles, you may come to realise that mathematics is beautiful, useful and fun.

THE TWO CULTURES

'Of course I've heard of Beethoven, but who is this guy Gauss?'

The 'Two Cultures', introduced by the British scientist and novelist C. P. Snow in an influential Rede Lecture in 1959, are still relevant today.

Ludwig van Beethoven and Carl Friedrich Gauss were at the height of their creativity in the early nineteenth century. Beethoven's music, often of great subtlety and intricacy, is accessible even to those of us with limited knowledge and understanding of it. Gauss, the master of mathematicians, produced results of singular genius, great utility and deep aesthetic appeal. But, although the brilliance and beauty of his work is recognised and admired by experts, it is hidden from most of us, requiring much background knowledge and technical facility for a true appreciation of it.

There is a stark contrast here. There are many parallels between music and mathematics: both are concerned with structure, symmetry and pattern; but while music is accessible to all, maths presents greater obstacles. Perhaps it's a left versus right brain issue. Music gets into the soul on a high-speed emotional autobahn, while maths has to follow a rational, step-by-step route. Music has instant appeal; maths takes time.

It is regrettable that public attitudes to mathematics are predominantly unsympathetic. The beauty of maths can be difficult to appreciate, and its significance in our lives is often underestimated. But mathematics is an essential thread in the fabric of modern society. We all benefit from the power of maths to model our world and facilitate technological advances. It is arguable that the work of Gauss has a greater impact on our daily lives than the magnificent creations of Beethoven.

In addition to utility and aesthetic appeal, maths has great recreational value, with many surprising and paradoxical results that are a source of amusement and delight. The goal of this book is to elucidate the beauty, utility and fun of mathematics by examining some of its many uses in modern society and to illustrate how it benefits our lives in so many ways.

YOU CAN DO MATHS ‖

Can we all do maths? Yes, we can! Everyone thinks mathematically all the time, even if they are not aware of it. We use simple arithmetic every day when we buy a newspaper, a cinema ticket or a pint of beer. But we also do more high-level mathematical reasoning all the time, unaware of the complexity of our thinking.

The central concerns of mathematics are not numbers, but patterns, structures, symmetries and connections. Take, for example, the Sudoku puzzles that appear daily in newspapers. The objective is to complete a 9 × 9 grid, starting from a few given numbers or clues, while ensuring that each row, each column and each 3 × 3 block contains all the digits from 1 to 9 once and only once. But the numerical values of the digits are irrelevant; what is important is that there are nine distinct symbols. They could be nine letters or nine shapes. It's the pattern that matters.

One Irish daily paper publishes these puzzles with the subscript *'There's no maths involved, simply use reasoning and logic!'* It seems that even the *idea* that something might be tainted by mathematics is enough to scare off potential solvers! Could you imagine the promotion of an exhibition in the National Gallery with the slogan *'No art involved, just painting and sculpture'*? If you can do Sudoku, you can do maths!

Whether you are discussing climate averages, studying graphs of house prices, worrying about inflation rates or working out the odds on the horses, you are thinking in mathematical mode. On a daily basis, you seek the best deal, the shortest route, the highest interest rate or the fastest way to get the job done with least effort. The principle of least action encapsulates the fundamental laws of nature in a simple rule. You are using similar reasoning in everyday life. Maximising, minimising, optimising: that's maths.

Maps and charts are ubiquitous in mathematics. They provide a means of representing complex reality in a simple, symbolic way. Subway maps are drastically simplified and deliberately distorted to emphasise what matters for travellers: continuity and connectivity. When you use a map of the London Underground, you are doing topology: that's maths.

Crossing a road, you observe oncoming traffic, estimate its speed and time to arrive, reckon the time needed to cross, compare the two and decide whether to walk or to wait. Estimating, reckoning, comparing: that's maths. Driving demands even more mathematical reasoning. You must constantly gauge closing speeds, accelerations, distances and times. Driverless cars are on the way: they use advanced mathematical algorithms and intensive computation. You can do that yourself in a flash.

Suppose you have the misfortune to fall ill. The doctor spells it out: the most effective treatment has severe side-effects; the alternative therapy is gentler but less efficacious; doing nothing has grave implications. A difficult choice must be made. You weigh up the risks and consequences of each course of action, rank them and choose the least-worst option. Weighing, balancing, ranking: that's maths.

Professional athletes can run 100 metres in ten seconds thanks to sustained, intensive training. Composers create symphonies after years of diligent study and practice. And professional mathematicians derive profound results through arduous application to their trade. You cannot solve technically intricate mathematical problems or prove arcane and abstruse theorems, but you can use logic and reasoning, and think like a mathematician. It is just a matter of degree.

INSTANT INFORMATION

Type a word into Google and a billion links appear in a flash. How is this done? How do computer search engines work, and why are they so good? *PageRank* (the name is a trademark of Google) is a method of measuring the popularity or importance of web pages. PageRank is a mathematical *algorithm*, or systematic procedure, at the heart of Google's search software. Named after Larry Page, a co-founder with Sergey Brin of Google, the PageRank of a web page estimates the probability that a person surfing at random will arrive at that page. Gary Trudeau, of *Doonesbury* fame, has described it as 'the Swiss Army knife of information retrieval'.

At school we solve simple problems like this: 6 apples and 3 pears cost €6; 3 apples and 4 pears cost €5; how much for an apple? This seems remote from practical use, and students may be forgiven for regarding it as pointless. Yet it is a simple example of simultaneous equations, a classical problem in linear algebra, which is at the heart of many modern technological developments. One of the most exciting recent applications is PageRank.

The PageRank computations form an enormous linear algebra problem, like the apples and pears problem but with billions of different kinds of fruit. The array of numbers that arises is called the 'Google matrix' and the task is to find a special string of numbers related to it, called the 'dominant eigenvector'. The solution can be implemented using a beautifully simple but subtle mathematical method that gives the PageRank scores of all the pages on the web.

The web can be represented as a huge network, with web pages indicated by dots and links drawn as lines joining the dots. Brin and Page used hyperlinks between web documents as the basis of PageRank. A link to a page is regarded as an indicator of popularity and importance, with the value of this link increasing with the popularity of the page linking to it. The key idea is that a web page is important if other important pages link to it.

Thus, PageRank is a popularity contest: it assigns a score to each page according to the number of links to that page and the score of each page linking to it. So it is *recursive*: the PageRank score depends on PageRank scores of other pages, so it must be calculated by an iterative process, cycling repeatedly through all the pages. At the beginning, all pages are given equal scores. After a few cycles, the scores converge rapidly to fixed values, which are the final PageRank values.

Google's computers or 'googlebots' are ceaselessly crawling the web and calculating the scores for billions of pages. Special programs called spiders are constantly updating indexes of page contents and links. When you enter a search word, these indexes are used to find the most relevant websites. Since these may number in the billions, they are ranked based on popularity and content. It is this ranking that uses ingenious mathematical techniques.

Efforts to manipulate or distort PageRank are becoming ever more subtle, and there is an ongoing cat-and-mouse game between search engine designers and spammers. Google penalises web operators who use schemes designed to artificially inflate their ranking. Thus, PageRank is just one of many factors that determine the search result you see on your screen. Still, it is a key factor, so those techniques you learned in school to find the price of apples and pears have a real-world application of great significance and value.

(The answer to the puzzle: apples cost 60 cents and pears cost 80 cents.)

NAPIER'S NIFTY RULES ‖

Spherical trigonometry is not in vogue. A century ago, a Tripos student at Cambridge might resolve half a dozen spherical triangles before breakfast. Today, even the basics of the subject are unknown to many students of mathematics. That is a pity, because there are many elegant and surprising results in spherical trigonometry. For example, two spherical triangles that are similar – having corresponding angles equal – have the same area. This contrasts sharply with the situation for plane geometry.

There is no denying the crucial importance of spherical trigonometry in astronomy and in the geosciences. A good memory is required to master all the fundamental results: the sine law, the cosine law for angles, the cosine law for sides, Gauss's formulae and many more. But we can get a long way with a few simple and easily remembered rules formulated by the inventor of logarithms.

The equation for a great circle involves the intersection of a plane and a sphere, an easy problem in three-dimensional Cartesian geometry. It is

$$\tan \phi = \tan \varepsilon \sin (\lambda - \lambda_0)$$

where λ and ϕ are longitude and latitude and the great circle crosses the equator through λ_0 at an angle ε. A more direct approach of showing this is possible: the formula for a great circle turns out to be one of Napier's Rules.

These rules are easy to state. Every spherical triangle has three angles and three sides. The sides are also expressed as angles, the angles they subtend at the centre of the sphere. For a sphere of unit radius, these angles (in radians) equal the lengths of the sides. *Napier's Rules apply to right-angled triangles*. Omitting the right angle, we write the remaining five angles in order on a pie diagram, but replace the three angles *not* adjacent to the right angle by their complements (their values subtracted from 90 degrees). If we select any three angles, we will always have a middle one and either two angles adjacent to it or two angles opposite to it. Then Napier's Rules are:

SIne of mIddle angle = Product of tAngents of Adjacent angles

SIne of mIddle angle = Product of cOsines of Opposite angles

With five choices for the middle angle and adjacent and opposite cases for each, there are ten rules in all. As a mnemonic, note the correspondences of the first vowels in key words, indicated in bold.

Napier's Rules apply only to right triangles, but we can often handle a general spherical triangle by dividing or extending it. Suppose we want to find out the great circle distance from Paris to Cairo, and we know the latitude and longitude of each city. The meridians from the North Pole to these cities, together with the great circle between them, form a spherical triangle for which we know two sides and the included angle. We can apply the cosine law for sides to get the great circle distance. But what if we have forgotten the cosine law? We can drop a perpendicular from Paris to the meridian through Cairo and apply Napier's Rules twice to find the inter-city distance (it turns out to be about 3,200 km).

John Napier (1550–1617), formulator of the rules, is best remembered as the inventor of logarithms. Also out of vogue today, his tables of logs enabled Johannes Kepler to analyse Tycho Brahe's observations and deduce the orbits of the planets. Napier also popularised the use of decimal fractions in arithmetic. But his work in mathematics was essentially recreational, for Napier was foremost a theologian. An ardent, even fanatical, Protestant, he regarded his commentary on the Book of Revelation as his best work. In *A Plaine Discovery of the Whole Revelation of St John*, he predicted that the apocalypse and the end of the world would occur in 1700.

Napier's book on logarithms contained his 'Rules of Circular Parts' of right spherical triangles. As we have seen, they are easily remembered and simple to apply. If you are ever marooned on a desert island and know the location, you can use them to work out how far you will have to swim home. I hope you make it.

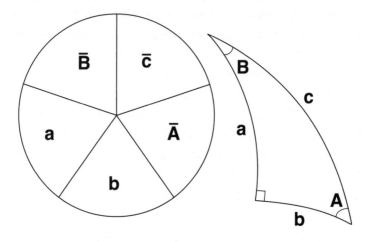

SPROUTOLOGY II

Sprouts is a simple and delightfully subtle pencil and paper game for two players. The game is set up by marking a number of spots on a page. Each player makes a move by drawing a curve that joins two spots, or that loops from a spot back to itself, without crossing any lines drawn earlier, and then marking a new spot on the curve. A maximum of three lines may link to a spot, and any spot with three lines is considered dead, since it plays no further role in the game. Sooner or later, no further moves are possible and the player who draws the last line wins the game.

Sprouts was devised by two Cambridge mathematicians, John Horton Conway and Michael Stewart Paterson, in 1967. It has an addictive appeal, and it immediately became a craze, being played in mathematics departments around the world. Despite the simple rules, the analysis of the game presents some challenges, and no general winning strategy is known. It is fairly easy to show that if there are n spots to start, the game will have at least $2n$ moves, and must end in at most $3n-1$. Thus, with 8 spots to start, there will be between 16 and 23 moves.

The mathematics of Sprouts, which we might call sproutology, involves topology, a form of geometry that considers continuity and connectedness but disregards distances and shapes. Topology is often called rubber-sheet geometry since a figure drawn on an elastic sheet retains its topological properties when the sheet is stretched but not torn. Sprouts is topological, since the precise positions of the spots is unimportant; it is only the pattern of connections between them that counts. The game exploits the Jordan curve theorem, which states that simple closed curves divide the plane into two regions. This apparently obvious result is actually quite difficult to prove.

The one-spot game of Sprouts is trivial: the first player must join the spot to itself and draw another spot; the second player then joins the two spots, winning the game. Games with a small number of starting spots have been fully investigated, and a pattern is evident: if the remainder when n is divided by 6 is 3, 4 or 5, the first player can force a win (assuming perfect play); otherwise, the second player has a winning strategy. This 'Sprouts conjecture' remains unproven.

For up to seven spots to start, Sprouts can be checked by hand, but for larger numbers of spots it rapidly becomes too complex and a computer analysis is required. Recently, Julien Lemoine and Simon Viennot analysed games with up to 47 spots, and their findings support the Sprouts conjecture. Of course, the existence of a winning strategy does not guarantee a win. Despite its elementary rules, Sprouts is surprisingly subtle, and prowess comes only with practice. You should start with a small number of spots, between five and 10, and gradually build up skill. But beware the addictive appeal of the game: you may well become a sproutaholic.

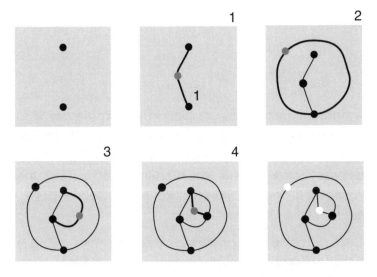

A sample game of Sprouts in which the second player wins after four moves.

WHY DON'T CLOUDS FALL DOWN?

A stone memorial was unveiled in 1995 in the tiny Sligo townland of Skreen to honour a great nineteenth-century mathematician and physicist who hailed from there. George Gabriel Stokes was born in Skreen in 1819, the youngest of seven children of Reverend Gabriel Stokes, Rector of the Church of Ireland.

George showed clear signs of brilliance from an early age, excelling at mathematics. After education in Skreen, Dublin and Bristol, he matriculated to Pembroke College, Cambridge, graduating in 1841 as Senior Wrangler; that is, gaining first place in the entire University of Cambridge in Part II of the Mathematical Tripos, the final mathematics examinations. Just eight years later he was appointed Lucasian Professor of Mathematics, a position that he held for over fifty years. This prestigious chair had earlier been held by Isaac Newton and more recently by Stephen Hawking.

Stokes's scientific interests were very broad, and he corresponded on a wide range of subjects with another giant of Victorian science, Belfast-born Lord Kelvin. A particular focus of his work was wave phenomena in various media. Some of his best-known research was on the theory of light waves. In this work, he obtained some major advances in the mathematical theory of diffraction and elucidated the phenomenon of fluorescence, the emission of light by a substance that has absorbed electromagnetic radiation. We benefit from this work through fluorescent lamps; these use electricity to excite mercury atoms, which then cause a phosphor coating to fluoresce, producing visible light.

Stokes investigated the internal friction of fluids, explaining how small droplets are suspended in the air and giving an answer to the age-old question asked by children: Why don't clouds fall down? His description of fluid viscosity was incorporated into the equations of fluid motion, now called the Navier–Stokes equations. These equations are of fundamental importance in all studies of fluid motion and are central to the study of turbulence, for modelling the oceans and for weather prediction and climate modelling.

In 1859 Stokes married Mary Susanna, daughter of Thomas Romney Robinson, Astronomer at Armagh Observatory. Robinson had an interest in the atmosphere and had invented the spinning cup anemometer for measuring wind speed. This interest must have influenced Stokes, who later developed an instrument called the Campbell–Stokes sunshine recorder.

In 1851, Stokes was elected a Fellow of the Royal Society. For thirty years he was secretary of the society and later served as its president. He was also an MP for a time, representing the University of Cambridge. But he never forgot his origins in Skreen, and returned to Sligo regularly for summer vacations. And in one of his heavily mathematical papers he wrote of 'the surf which breaks upon the western coasts as a result of storms out in the Atlantic', recalling the majestic rollers thundering in as he strolled as a boy along Dunmoran Strand near Skreen.

Stokes won many honours during his life, and his name is preserved in a large number of scientific contexts, including Stokes' Law (in fluid dynamics), Stokes' Theorem (in vector calculus), the Stokes shift (fluorescence), the Stokes phenomenon (in asymptotics) and many more.

PACKING ORANGES AND STACKING CANNONBALLS

Packing problems are concerned with storing objects as densely as possible in a container. Usually the goods and the container are of fixed shape and size. Many packing problems arise in the context of industrial packaging, storage and transport, in biological systems, in crystal structures and in carbon nanotubes, tiny molecular-scale pipes.

Packing problems illustrate the interplay between pure and applied mathematics. They arise in practical situations but are then generalised and studied in an abstract mathematical context. The general results then find application in new practical situations. A specific example of this interplay is the sphere-packing problem.

In 1600, the adventurer Walter Raleigh asked his mathematical adviser Thomas Harriot about the most efficient way of stacking cannonballs on a ship's deck. Harriot wrote to the famous astronomer Johannes Kepler, who formulated a conjecture that a so-called 'face-centred cubic' was the optimal arrangement.

Let's start with a simpler problem: How much of a table-top can you cover with non-overlapping €1 coins? Circular discs can be arranged quite densely in a plane. If they are set in a square formation, they cover about 79% of the surface. But a hexagonal arrangement, like a honeycomb, with each coin touching six others, covers over 90%; that's pretty good. Joseph-Louis Lagrange showed in 1773 that no regular arrangement of discs does better than this. But what about irregular arrangements? It took until 1940 to rule them out.

In three dimensions, we could start with a layer of spheres arranged in a hexagonal pattern like the coins, and then build up successive layers, placing spheres in the gaps left in the layer below. This is how grocers instinctively pile oranges, and gunners stack cannonballs. The geometry is a bit trickier than in two dimensions, but it is not too difficult to show that this arrangement gives a packing density of about 74%. The great Gauss showed that this is the best that can be done for a regular or lattice arrangement of spheres.

But again we ask: what about irregular arrangements? Is it not possible to find some exotic method of packing the spheres more densely? Kepler's Conjecture says 'No', and the problem has interested many great mathematicians in the intervening four hundred years. In 1900 David Hilbert listed 23 key problems for twentieth-century mathematicians, and the sphere-packing puzzle was part of his 18th problem.

In 1998 Thomas Hales announced a proof of Kepler's Conjecture. He broke the problem into a large number of special cases and attacked each one separately. But there were some 100,000 cases, each requiring heavy calculation, far beyond human capacity, so his proof depended in an essential way upon using a computer. After detailed review, Hales' work was finally published in 2005 in a 120-page paper in *Annals of Mathematics*. Thus, Kepler's Conjecture has become Hales' Theorem! Most mathematicians accept that the matter is resolved, but there remains some discomfort about reliance on computers to establish mathematical truth.

Why should we concern ourselves with a problem for which grocers and cannoneers knew the solution long ago? Well, in higher dimensions the corresponding problem has more intriguing aspects. It is a key result in data communication: to minimise transmission errors, we design codes that are based on maximising the packing density of hyper-spheres in high-dimensional spaces. So the apparently abstruse conjecture of Kepler has some eminently practical implications for our technological world.

MODELLING EPIDEMICS

The film *Contagion* painted a terrifying picture of the breakdown of society following a viral pandemic. The movie identified a key parameter, the basic reproduction number R-nought (R_0). This number measures how many new people catch the virus from each infected person, and is crucial in determining how fast an infection spreads.

In March 2003, an epidemic of severe acute respiratory syndrome (SARS) spread rapidly across the globe. The World Health Organisation issued a global alert after SARS had been detected in several countries. Since the spread of infections is greatly facilitated by international air travel, controls on movement can certainly be effective: with appropriate travel restrictions, the SARS epidemic was brought under control within a few months.

Epidemiological analysis and mathematical models are now essential tools in understanding and responding to infectious diseases such as SARS. Models range from simple systems of a few variables and equations to highly complex simulations with many millions of variables. A broad range of mathematics, both conventional techniques and methods emerging from current research, are involved. These include dynamical systems theory, statistics, network theory and computational science.

Public health authorities are faced with crucial questions: How many people will become infected? How many do we need to vaccinate to prevent an epidemic? How should we design programmes for prevention, control and treatment of outbreaks? The models allow us to quantify mortality rates, incubation periods, levels of threat and the timescale of epidemics. They can also predict the effectiveness of vaccination programmes and control policies, such as travel restrictions.

Parameters like transmission rates and basic reproduction numbers cannot be accurately estimated for a new infection until an outbreak actually occurs. But models can be used to study 'what if' scenarios to estimate the likely consequences of future epidemics or pandemics.

In a paper published in 1927, 'A Contribution to the Mathematical Theory of Epidemics', two scientists in Edinburgh, William Kermack and Anderson McKendrick, described a simple model with three variables, and three 'ordinary differential equations' that describe how infection levels change with time, which was successful in predicting the behaviour of some epidemics. Their model divided the population into three groups: susceptible, infected and recovered people, denoted S, I and R respectively. This SIR model simulates the growth and decline of an epidemic and can be used to predict level of infection, timescale and the total percentage of the population afflicted by the infection.

However, many important factors are omitted from the simple SIR model. The swine flu epidemic in Britain reached a peak in July 2009 and then declined rapidly and unexpectedly. The key factor not included in the model was the effect on the transmission rate of the school holidays, with contacts between children greatly reduced. The growth of the outbreak was interrupted, but an even larger peak was reached in October, after school had resumed. When these social mixing patterns were included, the model produced two peaks, in agreement with the observed development.

The statistician George Box, a pioneer in time series analysis, design of experiments and Bayesian inference, once remarked: 'All models are wrong, but some are useful.' All models of epidemics have limitations, and those using them must bear these in mind. Given the vagaries of human behaviour, prediction of the exact development of an infectious outbreak is never possible. Nevertheless, models provide valuable insights not available through any other means.

Future influenza pandemics are a matter of 'when' rather than 'if'. In planning for these, mathematical models will play an indispensable role.

A FALLING SLINKY ‖

If you drop a slinky from a hanging position, something very surprising happens. *The bottom remains completely motionless* until the top, collapsing downwards, coil upon coil, crashes into it.

How can this be so? We all know that anything with mass is subject to gravity, and this is certainly true of the lower coils of the slinky. But there's another force acting on them, the tension due to the stretching of the slinky. When hanging in an equilibrium position, these two forces, gravity and tension, balance exactly, so there is no movement.

When we let go of the top, the tension in the uppermost coils is relaxed and, since there is nothing to balance gravity, they start to fall. But this relaxation has to be transmitted or communicated down the slinky before gravity can pull the bottom downwards. This transmission takes time: the time for the 'message' to travel the length of the slinky depends on the ratio of the mass to the stiffness.

A slinky has large mass and small stiffness, so this time is relatively long, typically about half a second. But a freely falling object falls five metres in the first second. Moreover, the top coils of the slinky initially accelerate downwards even faster than in free fall, because the downward tension augments gravity. Thus, the slinky reaches a crunch point, where the top crashes into the bottom, before the signal of the release can reach it. You might say that *the bottom doesn't know what hit it*!

It is worthwhile playing with a real slinky to study this curious behaviour. If you put the slinky on a table, stretch it, hold one end steady and jerk the other end, you will see the signal propagating along the spring. But the best way to view the falling slinky is in slow motion.

There are several videos on YouTube illustrating falling slinkies, for example http://www.youtube.com/watch?v=uiyMuHuCFo4 .

A 'MERSENNERY' QUEST ‖

Prime numbers are of central importance in pure mathematics and also in a wide range of applications, most notably cryptography. The security of modern communication systems depends on their properties. Recall that a prime number is one that cannot be evenly divided by a smaller number. Thus, 2, 3 and 5 are primes, but 4 and 6 are not, since $4 = 2 \times 2$ and $6 = 2 \times 3$. Primes are the atoms of the number system: every whole number is a product of primes.

The search for patterns in the distribution of primes has occupied mathematicians for centuries. They appear to be randomly strewn among the whole numbers, but there are tantalising indications of structure. Often, a hint of a pattern emerges, only to evaporate upon further study. Thus, 31 is prime, as are 331, 3331, 33331 and the next three members of this sequence. But 333,333,331 is divisible by 17, and the pattern is broken.

In elementary algebra, we learn to solve quadratic equations. This corresponds to finding the zeros of a simple polynomial equation. The zeros of a more complicated function, called the zeta function, are intimately connected with the distribution of the prime numbers, but the location of all these zeros is enormously difficult. They are believed to satisfy a pattern first proposed in 1859 by Bernhard Riemann, but this has never been proved. The Riemann Hypothesis is widely regarded as the most important unsolved problem in mathematics. A proof would have far-reaching implications and whoever proves it will win lasting fame. They will also collect a $1 million prize from the Clay Mathematics Institute.

The frantic dash to find ever-larger prime numbers has been dominated in recent years by the Great Internet Mersenne Prime Search (GIMPS), a voluntary collaborative project involving a large number of personal computers. The record for the largest prime is broken on a regular basis. Almost all the recent examples have been found by GIMPS, and are numbers of a particular form called *Mersenne primes*, which are one less than a power of 2. As of June 2016, the largest known prime is obtained by multiplying 2 by itself 74,207,281 times and subtracting 1. With more than 22 million digits, it would fill many thousands of printed pages.

Mersenne numbers take their name from a seventeenth-century friar called Marin Mersenne. Born in France in 1588, Mersenne was a strong apologist for Galileo, whose scientific ideas challenged religious orthodoxy. Mersenne's main scientific work was in acoustics, but he is remembered today for his association with the Mersenne primes. He had contact with many mathematical luminaries, and provided a vital communication channel, corresponding with mathematicians, including René Descartes and Étienne Pascal, in many countries. Mersenne was, in essence, a one-man internet hub.

GIMPS has found the ten largest known prime numbers, and regularly smashes its own record. The project uses a search algorithm called the Lucas–Lehmer primality test, which is particularly suitable for finding Mersenne primes and is very efficient on binary computers. The test was originally developed by Édouard Lucas in the nineteenth century, and improved by Derrick Henry Lehmer in the 1930s.

For discovering a prime with more than 10 million decimal digits, GIMPS won a $100,000 prize and a Cooperative Computing Award from the Electronic Frontier Foundation (EFF). A prize of $150,000 is on offer from EFF for the first prime number found with at least 100 million decimal digits, and a further $250,000 for one with at least a billion digits. What are you waiting for?

SHACKLETON'S SPECTACULAR BOAT JOURNEY

A little mathematics goes a long, long way. Elementary geometry brought a small team of heroes 800 sea miles across the treacherous Southern Ocean, and resulted in 28 lives being saved.

For eight months, Ernest Shackleton's expedition ship *Endurance* had been carried along, ice-bound, until it was finally crushed and sank in October 1915. This put an end to the plans of the Irish-born explorer and his team of 28 men to cross the Antarctic continent. They salvaged three boats and made their way to Elephant Island, at the tip of the Antarctic Peninsula.

With five companions, Shackleton set out in one of the boats, a whaler called the *James Caird*, setting a course for South Georgia, some 800 nautical miles distant. With unceasing gales, the sea was tempestuous. Navigation depended on sightings taken with a sextant during rare appearances of the sun. Heavy rollers tossed the boat about, making it difficult to sight the horizon. The process was described by navigator Frank Worsley as 'a merry jest of guesswork'.

The strategy was to reach the latitude of South Georgia and let the westerly winds and currents carry the boat to the island. *Latitude* is measured by 'shooting the sun' with a sextant. The horizon and the lower limb of the sun are aligned in a split mirror, viewed through a telescope. The altitude of the sun can then be read from an indicator on the sextant arc. The geometry is straightforward: looking at the diagram below, we can see that the latitude θ is given by $\theta = 90° + \sigma - \alpha$ where α is the sun's altitude read from the sextant and σ is the latitude of the sun. This last depends on the date and time, and is given in the *Nautical Almanac*.

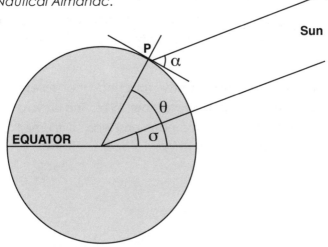

Angles used to calculate the latitude. Alpha (α) is the altitude of the sun measured with the sextant; sigma (σ) is the latitude of the sun, obtained from the Nautical Almanac; and theta (θ) is the latitude of point P.

To get the *longitude*, a clear shot of the sun at local noon is required. The navigator tracks the solar altitude to determine the exact time when the sun reaches its highest point. This is local apparent noon. The chronometer is set to Coordinated Universal Time (UTC or GMT). Since the earth rotates in 24 hours, the sun appears to move westwards 15 degrees in each hour. Thus, if the chronometer reads 15:00 GMT, local noon is three hours behind Greenwich and the longitude is 45° west.

After 17 days, Shackleton and his companions landed on the west coast of South Georgia, at about 54°. The voyage was a marvel of navigation, one of the greatest boat journeys ever accomplished. But the trouble was not over yet. Shackleton still had to cross the mountainous interior of the island to reach the whaling station at Stromness and arrange a rescue mission to relieve the men left behind on Elephant Island.

Ultimately, the entire party reached the safety of Punta Arenas, Chile in September 1916. The survival of Shackleton and all his companions was 'a triumph of hope and inspired leadership'.

WHERE IN THE WORLD? ‖

Most hill-walkers can recall an anxious time when, caught on a ridge between steep slopes, they were suddenly enshrouded by dense fog. A carefree ramble becomes a terrifying test of survival. The immediate question is 'Where exactly am I?' Map and compass are vital aids, but they cannot answer that question. A hand-held device about the size of a mobile phone can. How does it do that?

The Global Positioning System is a satellite-based navigation system, owned and operated by the US government, that provides information on location in all weathers, anywhere in the world. It is freely available to anyone with a GPS receiver, costing perhaps €100. The system comprises a constellation of between 24 and 32 satellites, orbiting at about 20,000 km above the earth. Each satellite carries a high-precision atomic clock, accurate to about one nanosecond. A nanosecond (ns) is one billionth of a second, the time it takes light to travel one foot.

To compute the position, the GPS receiver uses signals from several satellites, each including the precise time and location of the satellite. The satellites are synchronised so that the signals are transmitted at precisely the same instant. But they arrive at the GPS receiver at slightly different times. Using the known signal speed, the speed of light, the distance to each satellite is determined. These distances are then used to calculate the position of the receiver, using *trilateration*.

Trilateration determines position by using distances to known locations. This is in contrast to triangulation, which uses angles. For example, if you are 110 km from Athlone, you are somewhere on a circle of this radius centred at Athlone. If you are also 140 km from Belfast, you must be in Dublin or in Garrison, Fermanagh, the points where two circles intersect. Finally, if you are also 220 km from Cork, you can only be in Dublin. Three distances suffice for a unique location.

In three-dimensional space, spheres replace circles and four are needed, so the GPS receiver uses signals from four satellites. This provides distances from four known locations, sufficient to pin down the position of the receiver. GPS receivers available today give location to an accuracy of about ten metres. This position may be plotted on a background map or given as latitude and longitude or a National Grid reference.

Navigation is just one of the many civilian and military applications of GPS. The system is vital for search and rescue, for vehicle tracking, for map-making and surveying and for detecting movements in the earth's crust. Monitoring the movements of elephants in Africa is one among many other applications. Satnav is considered so essential that the European Union is developing a GPS system called Galileo. As of June 2016 there were 14 of 30 satellites in orbit and the system should be fully operational by 2019.

GPS is a striking example of the practical importance of Einstein's relativity theory. Special relativity implies that a moving clock ticks slowly relative to a stationary one, so for an observer on earth, the satellite clocks lose about 7,000 ns (7 microseconds) each day. But general relativity says that these clocks should go about 45,000 ns *faster*, because the earth's gravitational pull is weaker higher up. The net effect is a speed-up of about 38,000 ns per day. To avoid cumbersome corrections, the clocks are reset before launch to compensate for relativistic effects. Without this, GPS would be useless for navigation!

The Global Positioning System is a remarkable synthesis of old and new. It involves high-tech engineering and complex relativistic physics to enable it to function, but the mathematics used to determine location is simple, being a straightforward application of the geometry of circles and spheres developed in ancient Greece.

SRINIVASA RAMANUJAN

Srinivasa Ramanujan, one of the greatest mathematical geniuses ever to emerge from India, was born in 1887 into a poor Brahmin family. Ramanujan had limited formal education but was consumed by his passion for mathematics. He neglected all other subjects and failed the entrance exam for the University of Madras. However, he continued his mathematical research with intensity.

In 1913, Ramanujan wrote to G. H. Hardy, the leading mathematician in Britain, enclosing some of his results. Hardy examined them and concluded that they 'could only be written down by a mathematician of the highest class'. Thus began one of the most successful mathematical collaborations of all time. For five years, Ramanujan worked with Hardy in Cambridge, publishing many papers of great richness and originality. In 1918 he was elected a Fellow of the Royal Society.

Ramanujan returned to India in 1919, but lived for only one more year. Shortly before his death, aged only 32, Ramanujan wrote a last letter to Hardy in which he introduced 17 completely new and strange power series that he called 'mock theta functions'.

In 1976 the American mathematician George Andrews was looking through some papers in the Wren Library in Cambridge and recognised Ramanujan's handwriting. What he found, now known as the 'lost notebook', contains many remarkable results, including Ramanujan's results on the mysterious mock theta functions.

Andrews' discovery opened up a vast new landscape. The results were of stunning novelty, representing what many regard as Ramanujan's deepest work. The finding of the lost notebook has been compared to finding a manuscript of Beethoven's Tenth Symphony. The consequences have been profound, for both pure mathematics and theoretical physics.

Ramanujan gave no clue as to how he had discovered the mock theta functions. An intrinsic meaning of them has eluded mathematicians until very recently. Sander Zwegers, a lecturer at UCD until he moved to Cologne in 2011, finally explained how they fit into a broader context. Zwegers' 2002 PhD thesis was groundbreaking, and has led to numerous publications and international conferences.

The breakthrough in our understanding is having an impact on many aspects of mathematics and physics. In pure mathematics, the results have been applied to graph theory, group theory and differential topology. In physics, there are applications in particle physics, statistical mechanics and cosmology. In particular, Ramanujan's functions have proved valuable for calculating the entropy of black holes.

Ramanujan's startlingly brilliant and innovative research paved the way for many major breakthroughs in number theory over the past century. Mathematician and theoretical physicist Freeman Dyson spoke of 'a grand synthesis still to be discovered', and he speculated about applications of the mock theta functions to string theory. This is an indication of the prescience and genius of Ramanujan's work, confirming Hardy's description of him as having 'profound and invincible originality'.

SHARING A PINT II

Four friends, exhausted after a long hike, stagger into a pub to slake their thirst. But, pooling their funds, they have enough money for only one pint.

Annie drinks first, until the surface of the beer is halfway down the side (Fig. 1 (A)). Then *Barry* drinks until the surface touches the bottom corner (B). *Cathy* then takes a sup, leaving the level as in (C), with the surface through the centre of the bottom. Finally, *Danny* empties the glass.

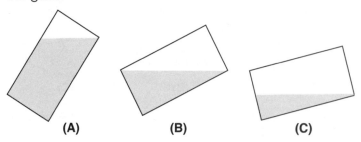

(A)　　　　　**(B)**　　　　　**(C)**

Figure 1

Question: Do all four friends drink the same amount? If not, who gets most and who gets least?

By symmetry, Annie has drunk half of the top half of the glass. So she has consumed 25% of the beer. Again by symmetry, Barry has left exactly 50% of the beer in the glass, so he has swallowed 25%. So far so good.

But Cathy has left beer forming a less regular shape: the liquid remaining in (C) is in the shape of an ungula, the volume formed by a plane slicing a cylinder and passing through the centre of the base. We have to calculate the volume of the ungula to see how much beer is left for Danny.

Ungula means hoof, and a section of a cylinder or cone cut off by a plane oblique to the base is so called because it resembles a horse's hoof. The shape is shown in Figure 2 below. Its volume can be calculated by a mathematical operation known as integration.

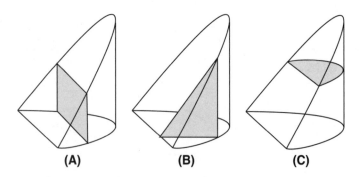

(A)   **(B)**   **(C)**

Figure 2 Cross-sections of an ungula perpendicular to (A) the x axis, (B) the y axis and (C) the z axis.

The three panels in Figure 2 show cross-sections of the ungula perpendicular to the x, y and z axes. They are respectively a rectangle, a triangle and a segment, and the volume is obtained by integration along the relevant axis. So, schematically, we can write the volume as

$$V = \int (\text{Rectangle}) \, dx = \int (\text{Triangle}) \, dy = \int (\text{Segment}) \, dz$$

where the symbol $\int$ denotes an integral, or sum over all the relevant shapes. Naturally, all three yield the same result, $V = (2/3)r^2h$ where r is the radius and h the height.

Now, the volume of the cylinder is πr^2h, so the fraction left for Danny is $2/(3\pi)$ or about 21%. Thus, while Annie and Barry drank 25% of the beer, Cathy must have drunk about 29%, leaving Danny short.

There is something remarkable about the volume of the ungula: it does not involve π, even though one of the surfaces is curved. **Where has π gone?**

More remarkable still is that Archimedes showed that the volume of the ungula is one-sixth that of the surrounding cube or block. The volume is $(2/3)r^2h$, and the volume of the rectangular box containing the cylinder is $2r \times 2r \times h = 4r^2h$, so indeed the ratio is $1/6$.

Archimedes used the triangular cross-section, integrating in the y direction. His reasoning was not some crude approximation, but a true application of the method of integral calculus.

‖ PONS ASINORUM

The fifth proposition in Book I of Euclid's *Elements* states that the two base angles of an isosceles triangle are equal (in the figure on page 45, angles B and C; an isosceles triangle is one having two equal sides). For centuries, this result has been known as the Pons Asinorum, or Bridge of Asses, apparently a metaphor for a problem that separates bright sparks from dunces.

Euclid proves the proposition by extending the sides AB and AC and drawing lines to form additional triangles. His proof is quite complicated. A simpler approach, popular for a hundred years or so, is to draw the line that bisects the apex angle A, splitting the triangle into two parts, which are then shown to be congruent, or equal in all respects. This requires use of an earlier result, Euclid's proposition I.4, which says that two triangles are congruent if they have two sides and the included angle equal.

Around 1960 another proof appeared, allegedly discovered by a computer. It ingeniously compared the triangle ABC to its mirror image ACB (see figure) and used Proposition I.4 to show that they are congruent, whence angle B equals angle C. This 'new' proof is intriguing in that it treats the triangle and its mirror image as separate for purposes of deduction but identical for purposes of conclusion.

When first published, this proof was considered to provide a convincing demonstration that computers can be creative. It was frequently cited as evidence of artificial intelligence (AI), for example in Douglas Hofstadter's remarkable book *Gödel, Escher, Bach: An Eternal Golden Braid*. But Michael Deakin of Monash University, Australia has investigated the matter. He reports an interview in 1981 in the *New Yorker*, in which AI guru Marvin Minsky of MIT stated that he produced the proof himself 'by hand simulation of what a machine might do'.

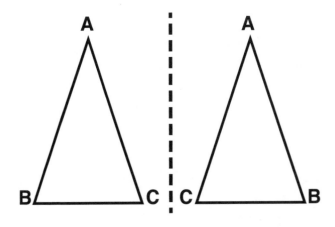

Amazingly, the ingenious proof was first discovered by the last great Greek geometer, Pappus of Alexandria, working around AD 320. It was derided by the nineteenth-century Oxford mathematician C. L. Dodgson, who imagined the reaction of Euclid: 'Surely that has too much of the Irish bull about it.' Dodgson was none other than Lewis Carroll, author of *Alice's Adventures in Wonderland*.

But what of the 'standard proof' using the bisector of the apex angle? Deakin points out that the reasoning in *A School Geometry*, a book by Hall and Stevens that some of us slaved over long ago, is circular. The proof of Pappus, rediscovered by Minsky and wrongly attributed to a computer, is certainly elegant. But perhaps it is safest to stick with Euclid's original proof. At least one child produced the Pappus proof in an examination and was marked wrong for it.

LOST AND FOUND: THE SECRETS OF ARCHIMEDES

Archimedes of Syracuse was the greatest mathematician of antiquity. He was also a brilliant physicist, engineer and astronomer, famed for founding hydrostatics, for formulating the law of the lever, for designing the helical pump that bears his name, for designing engines of war, and for much more. Generations of children have learned how, upon discovering a way to assay King Hieron's crown, Archimedes ran naked through the streets crying 'Eureka!'

Archimedes estimated the value of π, the ratio of the circumference to the diameter of a circle, to remarkable accuracy, using polygons of 96 sides within and around a circle. And he found the volume of a sphere, showing that it is two-thirds of the volume of the smallest cylinder in which it is contained. He asked that an image of a sphere within a cylinder be inscribed on his tombstone. Centuries later, the Roman orator Cicero found such a carving on a grave in Syracuse.

Many of Archimedes' writings are lost, known to us only through references made to them by later writers. Other works have reached us by a circuitous route: they were translated into Arabic in the ninth century, and from Arabic into Latin during the Renaissance. But some of Archimedes' most important work remained hidden from us until the remarkable discovery of the Archimedes Palimpsest.

Palimpsests were works written on parchment that had been scraped clean of earlier writing. This was common practice in the Middle Ages because vellum was very expensive. In 1906 a prayer book written in the thirteenth century came to light in Constantinople. Upon close examination by the Danish philologist Johan Heiberg, the incompletely erased work underlying the text was recognised as a tenth-century copy of several works of Archimedes, which had been thought to have been lost for ever.

The palimpsest is the only source we have of *The Method of Mechanical Theorems*, in which Archimedes uses infinitesimal quantities to calculate the volumes of various bodies. This method foreshadowed integral calculus, invented independently by Newton and Leibniz nearly two thousand years later.

The Archimedes Palimpsest is the earliest extant manuscript of Archimedes' work; it includes copies of the geometric diagrams that he drew in the sand in the third century BC. It contains several treatises by Archimedes, including *The Method*. The palimpsest was bought at auction in New York in 1998 for $2 million. The German magazine *Der Spiegel* reported that the purchaser was Jeff Bezos, founder of Amazon.com.

The palimpsest has been intensively analysed over the past ten years, using advanced imaging methods. A recent exhibition at the Walters Art Museum in Baltimore, 'Lost and Found: The Secrets of Archimedes', was devoted to the analysis of the palimpsest and to the outcome of the project to study it. All the images and translations are freely available on the Archimedes Palimpsest website (www.archimedespalimpsest.org), providing a treasure trove for scholars of Greek mathematics.

A page of the palimpsest showing older and more recent writing in orthogonal directions (from www.archimedespalimpsest.org).

SUBTERRANEAN TOPOLOGY

The London Underground map is a paragon of design excellence. If you know where you are and where you want to go, it shows you how to get there. But as a map of London it is inaccurate in almost all respects. The beauty of the design, originated by Harry Beck in 1931, is that the key information is kept, and everything else is stripped away.

The Tube map is what mathematicians call a graph. The stations are the *vertices* and the train lines joining them are the *edges*. Interchanges are shown where different lines connect. Distances and directions are distorted in the interests of clarity and simplicity. One of the earliest such graphs was drawn by the renowned Swiss mathematician Leonhard Euler. Euler solved a puzzle called 'The Seven Bridges of Königsberg' by drastically simplifying a map of that city. This made it clear that it is impossible to find a route crossing all seven bridges without recrossing any of them.

Graph theory is a branch of *topology*, the branch of mathematics dealing with continuity and connectivity.

Topology is concerned with properties that remain unchanged under continuous deformations, such as stretching or bending, but not cutting or gluing.

Topology is often called rubber sheet geometry. If a figure such as a triangle is drawn on a sheet of rubber and the sheet is stretched, certain things change but others remain unaltered. For example, the lengths of the sides are changed, but points inside the figure remain inside and points outside remain outside.

In three dimensions, a cube made of plasticine may be distorted continuously into a ball without tearing it, so a cube and a ball are topologically equivalent. In contrast, to make a bagel, or a doughnut with a hole, a ball of plasticine must be torn at some point. So a ball and a bagel are not equivalent.

The formal way of showing that two sets are topologically equivalent is to establish a correspondence or mapping between the two sets, such that nearby points in one are mapped to nearby points in the other. If such a correspondence – called a *homeomorphism* – exists, the two sets are topologically equivalent.

In the familiar school geometry of Euclid, we have straight lines, fixed distances between points and rigid shapes such as triangles. Since topological deformations sacrifice all these, is there anything useful left? Yes: while the London Tube map distorts distances, it preserves the order of stations and the connections between lines, so the traveller knows where to get on and off and where to change trains. It is this topological information that is critical; precise distances are of secondary importance.

The Tube map might be 'corrected' by drawing it on a sheet of rubber and delicately stretching it in places, gradually but continuously, until the stations are all in the correct positions. Or it might be further distorted until the Circle Line became a true circle. But the remarkable success and longevity of the map proves that Harry Beck got it just right all those years ago.

THE EARTH'S VAST ORB ‖

The shape of the earth has been a topic of great interest to savants for millennia. It is an over-simplification to say that the ancients believed the world to be flat. Just to watch a ship appear or disappear over the horizon, or to climb a mountain and notice the changing perspective of a distant island, is enough to provide a hint about the curved nature of the planet. But the prevalent view of earth was one of a vast flat plane surface; only a few people had greater insight.

Eratosthenes, a Greek mathematician, astronomer and geographer, went further than others and made an estimate of the earth's circumference that is close to the true value. His method, simple but clever, demonstrates the power of geometric reasoning. Eratosthenes knew that in midsummer the noonday sun was overhead in the city of Syene, modern-day Aswan, on the Tropic of Cancer. Observers there had noticed that at midday the sun's rays reached the bottom of a deep well.

But in Alexandria, when Eratosthenes measured the sun's angle relative to the zenith he obtained a value of about 7 degrees, or 1/50th of a circle. So, if the earth were a sphere, its circumference must be 50 times the distance from Aswan to Alexandria. Eratosthenes knew this distance to be about 800 km (in modern units), so he deduced a value close to 40,000 km for the earth's circumference, a remarkably accurate result.

Eratosthenes was librarian at the Great Library in Alexandria. He devised a system of latitude and longitude and made the first map of the known world with parallels and meridians. He is sometimes called the Father of Geography because he in effect invented the discipline and was the first person to use the term 'geography'.

Eratosthenes was a contemporary and friend of Archimedes. Eratosthenes himself was no mean mathematician. In addition to his use of geometry for map-making and measuring the size of the earth, he devised a simple method, or algorithm, for finding prime numbers, those numbers that cannot be evenly divided into smaller parts. List the natural numbers, 1, 2, 3, ..., up to some limit, say 500; then strike out every second number following 2, every third number after 3, every fifth after 5 and so on. Finally, only the prime numbers, 2, 3, 5, 7, 11, ..., remain. The procedure, the simplest way of listing small prime numbers, is known as the 'sieve of Eratosthenes'.

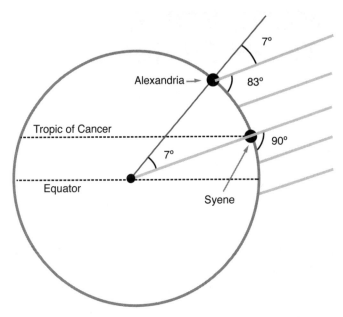

The angle of the noonday sun measured in midsummer at Alexandria and Syene (Aswan).

‖ MORE EQUAL THAN OTHERS

In his scientific bestseller, *A Brief History of Time*, Stephen Hawking remarked that every equation he included would halve sales of the book, so he put only one in it, Einstein's equation relating mass and energy, $E = mc^2$. There is no doubt that mathematical equations strike terror in the hearts of many readers. This is regrettable because equations are really just concise expressions of precise statements. They are actually quite user-friendly and more to be loved than feared.

An equation indicates that whatever is to the left side of the 'equals' sign has the same value as whatever is to the right. For Einstein's equation, E is the energy and it is equal to the mass multiplied by the square of the speed of light. In ancient times, equalities were expressed in verbal terms like this. It was Robert Recorde, a Welsh-born mathematician, who introduced the symbol = for 'equals'. In his book *The Whetstone of Witte*, written in 1557, Recorde wrote that he chose this symbol consisting of two parallel lines 'bicause no 2 thynges can be moare equalle'.

The first equation to appear in symbolic form was in Recorde's book, and was

$$14x + 15 = 71.$$

The quantity x is called the unknown (although early writers on mathematics called it 'the thing') and the equation states that if we take 14 times x and add 15 we get 71. How might such an equation arise? Suppose you need a hammer and some nails. A hammer costs €15 and a packet of nails is €14. If you buy a hammer and x packets of nails, the total cost is $14x + 15$, the left side of Recorde's equation. If you have just €71 to spend, how many packets of nails can you buy? The answer is the solution x of the equation.

Recorde explained the transformations that can be made to an equation to 'solve' it – that is, to find the unknown quantity x. In the present case, you can subtract 15 from each side to get a new equation $14x = 56$ and then divide both sides by 14 to get another one, $x = 4$. This is the solution, and you can afford four packets of nails.

Recorde is credited with introducing algebra into England with his book. But he had other talents in addition to mathematics. He was physician to Edward VI and Mary I, and in 1551 was appointed Surveyor of the Mines and Monies of Ireland. Alas, he ended his days in prison, for reasons that are unclear. Perhaps he became embroiled in a religious controversy, or ensnared in some political intrigue. Or perhaps some of the 'Monies of Ireland' went astray.

MATHS AND CAT SCANS

Many lives are saved each year through a synergistic combination of engineering, computing, physics, medical science and mathematics. This combination is CT imaging, or 'computed tomography', which is now an essential tool for medical diagnosis.

The story began in 1895, when Wilhelm Röntgen made the first radiograph using what he called X-rays. These high-energy electromagnetic beams can penetrate body tissues where light cannot reach. Internal organs can be examined non-invasively and abnormalities located with precision. For his trailblazing work, Röntgen was awarded the first Nobel Prize in Physics in 1901.

The power and utility of X-ray imaging has been greatly expanded by combining X-rays with computer systems to generate three-dimensional images of organs of the body. The diagnostic equipment used to do this is called a CT scanner (or CAT scanner). The word 'tomography' comes from the Greek *tomos*, meaning slice, and a CT scan is made by combining X-ray images of cross-sections or slices through the body. From these, a 3-D representation of internal organs can be built up.

Radiologists can use CT scans to examine all the major parts of the body, including the abdomen, chest, heart and head. In a CT scan, multiple X-ray images are taken from different directions. The X-ray data are then fed into a tomographic reconstruction program to be processed by a computer. The image reconstruction problem is essentially a mathematical procedure.

The tissue structure is deduced from the X-rays using a technique first devised by an Austrian mathematician, Johann Radon. He was motivated by purely theoretical interests when, in 1917, he developed the operation now known as the Radon transform. He could not have anticipated the great utility of his work in the practical context of CT. Reconstruction techniques have grown in complexity, but are still founded on Radon's work.

As they pass through the body, X-rays are absorbed to different degrees by body tissues of different optical density. The total attenuation, or dampening, is expressed as a 'line integral', the sum of the absorptions along the path of the X-ray beam. The more tissue along the path, and the denser that tissue, the less intense the beam becomes. The challenge is to determine the patterns of normal and abnormal tissue from the outgoing X-rays.

If the X-ray patterns were uncorrupted, the mathematical conversion to 3-D images would be straightforward. In reality, there is always noise present, and this introduces difficulties: Radon's 'inverse transform' is very unstable and error-prone, so a stable modification of the method, known as 'filtered back-projection', is used. More accurate algorithms have been developed in recent years, and research in this field is continuing.

Applications of tomography are not confined to medicine. The technique is also used in non-destructive materials testing, both in large-scale engineering and in the manufacture of microchips. It is also used to compute ozone concentrations in the atmosphere from satellite data. In addition to CT, there are numerous other volume-imaging techniques. Electron tomography uses a beam of electrons in place of the X-rays, ocean acoustic tomography uses sound waves, and seismic tomography analyses waves generated by earth movements to understand geological structures. All involve intricate mathematical processing to produce useful images from raw data.

BAYES RULES OK   ‖

In May 2009, en route from Rio de Janeiro to Paris, Air France flight AF447 crashed into the Atlantic Ocean. Bayesian analysis played a crucial role in the location of the flight recorders and the recovery of the bodies of passengers and crew. What is Bayesian analysis?

Classical and Bayesian statistics interpret probability in different ways. To a classical statistician, or frequentist, probability is the relative frequency of an event. If it occurs on average 3 out of 4 times, he or she will assign to it a probability of 3/4.

For Bayesians, probability is a subjective way to quantify belief, or degree of certainty, based on incomplete information. All probability is conditional and subject to change when more data emerge. If a Bayesian assigns a probability of 3/4, he or she should be willing to offer odds of 3 to 1 on a bet.

Frequentists find it impossible to draw conclusions about once-off events. By using prior knowledge, Bayesian analysis can deal with individual incidents. It can answer questions about events that have never occurred, such as the risk of an asteroid smashing into the earth or the chance of a major international war breaking out over the next ten years.

The danger of a major accident for the Challenger space shuttle was estimated by a Bayesian analysis in 1983 as 1 in 35. The official NASA estimate at the time was an incredible 1 in 100,000. In January 1986, during the 25th launch, the Challenger exploded, killing all seven crew members.

Bayes' Rule transformed probability from a statement of relative frequency into a measure of informed belief. In its simplest form, the rule, devised in the 1740s by the Reverend Thomas Bayes, tells us how to calculate an updated assessment of probability in the light of new evidence. We start with a prior degree of certainty. New data then make this more or less likely.

An advantage of Bayesian analysis is that it answers the questions that scientists are likely to ask. But, despite spectacular successes, Bayesian methods have been the focus of major controversy and their acceptance has been slow and tortuous. Through most of the twentieth century, the academic community eschewed Bayesian ideas and derided practitioners who applied them.

The nub of the controversy was that the probability using Bayesian methods depends on prior opinion. When data are scarce, this yields a subjective rather than an objective assessment. When information is in short supply, subjective opinions may differ widely.

The controversy was long and often bitter, with aspects of a religious war. The opponents vilified each other, generating great hostility between the two camps. The war is now over: frequentists and Bayesians both recognise that the two approaches have value in different circumstances.

Today, Bayesian analysis plays a crucial role in computer science, artificial intelligence, machine learning and language translation. Applications include search and rescue operations, like the recovery of the Air France flight AF447 black box, risk analysis, image enhancement, face recognition, medical diagnosis, setting insurance rates, filtering spam email and much more. Bayesian inference is likely to find many new applications over the coming decades.

PYTHAGORAS GOES GLOBAL

*About binomial theorem I'm teeming with a
  lot o' news,
With many cheerful facts about the square
  of the hypotenuse.*

('I Am the Very Model of a Modern Major-
General', from Gilbert and Sullivan, *The Pirates
of Penzance*)

Spherical trigonometry has all the qualities we expect of the best mathematics: it is beautiful, useful and fun. It played an enormously important role in scientific development for thousands of years, from ancient Greece through India, the Islamic Enlightenment and the Renaissance, to more modern times. It was crucial for astronomy, and essential for global navigation. Yet it has fallen out of fashion, and is almost completely ignored in modern education.

PYTHAGORAS ON THE SPHERE

Napier's Nifty Rules (page 11) give ten relationships between the angles and sides of right-angled triangles on the sphere. (Recall that the sides of the triangle are expressed in terms of the angles they subtend at the centre; the radius of the sphere is taken to be unity.) One – and only one – of Napier's Rules relates the three sides of the right-angled triangle. We take the right angle to be C, and the side opposite to it to be c. Then the rule is

$$\cos c = \cos a \cos b$$

This beautifully simple equation may not appear familiar but, believe it or not, this is just the spherical form of Pythagoras' Theorem!

Let us consider a small triangle, so all the sides a, b and c are small quantities. Then we may replace the cosine functions by their first few terms:

$$\cos a \approx (1 - \tfrac{1}{2} a^2) \,, \cos b \approx (1 - \tfrac{1}{2} b^2) \,, \cos c \approx (1 - \tfrac{1}{2} c^2) \,.$$

To second-order accuracy, we can write the equation $\cos c = \cos a \cos b$ as

$$c^2 = a^2 + b^2,$$

the usual form of Pythagoras' Theorem that we all know and love.

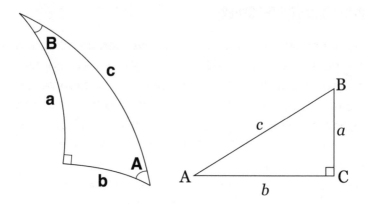

Left: A spherical right-angled triangle with cos c = cos a cos b
Right: A plane right-angled triangle with c² = a² + b²

DOZENAL DIGITS: FROM DIX TO DOUZE ‖

How many fingers has Mickey Mouse? He has three fingers and a thumb on each hand, so eight in all. Thus we may expect Mickey to reckon in octal numbers, with base eight. We use decimals, with ten symbols from 0 to 9 for the smallest numbers and larger numbers denoted by several digits, whose position is significant. Thus, 47 means four tens plus seven units.

But the base ten is divisible only by 2 and 5. There are advantages to having a highly composite base – one with many divisors. The Sumerians and Babylonians used base 60, divisible by 2, 3, 4, 5, 6, 10, 12, 15, 20 and 30. We still use remnants of this sexagesimal system in reckoning time and measuring angles, but base 60 is uncomfortably large for general use.

The duodecimal, or *dozenal*, system with base twelve has been proposed because 12 is divisible by 2, 3, 4 and 6. We need two extra symbols for the numbers ten and eleven, which are less than the base. Let's write them as X and E and call them *dek* and *el*. Then twelve is written 10 and called do (pronounced *doh* and short for a dozen). The system continues with 11, 12, ... 1X, 1E and 20 or do one, do two, ... do dek, do el and twodo.

Then, jumping in twelves, threedo, fourdo, up to eldo and gro. This gro is short for gross or twelve twelves, written 100. Twelve gro is one mo (twelve cubed or 1728 in decimal). So the decimal number 47 becomes 3E, threedo el or three twelves and el units. And we are currently (in 2016) in the year 1200, or mo twogro.

What advantages has the dozenal system? For one thing, multiplication tables are substantially simpler in dozenal. And many small fractions (one-quarter, one-third, three-quarters, etc.) have a simpler form in this system. So why don't we move from dix to douze? The Dozenal Societies of America and of Britain would favour this. We already have twelve months in a year and twice twelve hours in a day. But a number base change would be seriously disruptive, causing unimaginable confusion.

Computers convert numbers to binary form, using only zeros and ones, and convert the answer back to decimal before presenting it. We are generally oblivious to what goes on under the bonnet, and unconcerned about it.

The chance of the Dozenal Societies persuading us to change to base twelve is about the same as the likelihood of Mickey Mouse converting us to octal. But hold on: How many toes has Mickey got? In the notorious phrase beloved of maths book writers, this is left as an exercise for the student.

HOW LEOPARDS GET THEIR SPOTS ‖

Mathematical models enable us to understand many features of a growing embryo. For example, the patterns of hair colour that give leopards their spots and tigers their stripes can be produced by solving a mathematical equation with different inputs.

The information to form a fully grown animal is encoded in its DNA, so there is a lot of data in a single cell. But there are only about three billion base pairs in DNA and tens of trillions of cells in the body. So minute details like the twists and whorls of a fingerprint cannot be predetermined. Rather, they emerge during embryonic growth as a result of conditions determined by the DNA, following the basic laws of physics and chemistry.

Alan Turing is famous for cracking the Enigma code during World War II, but he was a polymath and worked on many other problems. In 1952, Turing published a paper, 'The chemical basis of morphogenesis', presenting a mechanism of pattern formation. He developed a theory of how the chemistry in the cell influences factors like hair colour.

Turing's model included two chemical processes: *reaction*, in which chemicals interact to produce different substances; and *diffusion*, in which local concentrations spread out over time. Suppose we have two chemicals, A and B, called morphogens, with A triggering hair colouring and B not doing so. Now suppose that A is a catalyst, stimulating production of further morphogens, whereas B suppresses production of them. Thus, A is called an activator and B an inhibitor.

Where A is abundant, the hair is black; where B is dominant, it is white. Now comes Turing's crucial assumption: the inhibitor B spreads out faster than the activator A. So B fills a circular region surrounding the initial concentration, forming a barrier where concentration of A is reduced. The result is an isolated spot of black hair where A is plentiful.

What is going on here is a competition between the reaction and diffusion processes. Many reaction–diffusion models have been proposed. The resulting patterns depend on the reaction rates and diffusion coefficients, and a wide range of geometrical patterns of hair colouring can result from this mechanism.

The figure opposite shows the concentration of chemical A for varying strengths of reaction and diffusion. High values of A are shaded black since hair colouring in these regions is expected to be black. For strong diffusion, the regions are large and striped like a tiger. For weak diffusion, the black hair is confined to spots like the coat of a cheetah.

Many other patterns can be generated by varying the parameters. Thin stripes, like those on an angel fish, or thick stripes, like those of a zebra, can be generated, and clusters of spots found on a leopard's coat can be produced. Biological systems are hugely complex, and simple mathematical models are valuable for elucidating key factors and predicting specific aspects of behaviour.

Concentration of constituent A at equilibrium predicted by the Schnakenberg equations. Black indicates high concentrations. Upper left: $\gamma = 100$. Upper right: $\gamma = 400$. Lower left: $\gamma = 1600$. Lower right: $\gamma = 6400$.

MONSTER SYMMETRY AND THE FORCES OF NATURE

In the arts, symmetry is intimately associated with aesthetic appeal. In science, it provides insight into the properties of physical systems. In 1918, the German mathematician Emmy Noether proved a powerful result relating symmetry and conserved quantities – quantities, such as energy, that remain constant. Her theorem is of central importance in classical and quantum mechanics. The branch of mathematics that formally considers symmetry is called *group theory*.

A group is a set of elements (they may be numbers or operations) together with a way of combining them: effectively, any two group elements may be multiplied to yield a third. For example, a square may be transformed and yet appear unchanged: it may be rotated through 90, 180, 270 or 360 degrees; it may be reflected about a vertical or horizontal line through the centre, or about either of its diagonals. These eight operations form a group: any two combine to yield another. It is called the octic group, and it is a beautiful object, elucidating many aspects of group theory.

Groups provide a powerful, unifying mathematical concept. As the renowned mathematics writer Eric Temple Bell observed, 'wherever groups disclosed themselves, or could be introduced, simplicity crystallised out of comparative chaos.'

Symmetries and groups are inextricably intertwined. Symmetries come in two flavours. There are discrete symmetries such as the bilateral symmetry of a human face, and continuous symmetries like the unvarying form of a sphere as it is rotated. Discrete symmetries are associated with finite groups – those having a finite number of operations. Continuous symmetries correspond to infinite groups.

There is a special class of groups that, like the elements of the periodic table, or the prime numbers, cannot be broken down into smaller pieces. These are the simple groups, sometimes called the atoms of symmetry. The octic group of symmetries of a square is not simple, but it can be decomposed into simple subgroups.

One of the great mathematical challenges of recent decades has been the classification of all finite simple groups; that is, the identification, arrangement and construction of all the possible cases. This programme, called by one mathematician the 'Thirty Years War', was finally completed in 2004. The programme gave rise to several surprises. In addition to 18 series of classical groups, there emerged 26 exceptional cases, called *sporadic groups*. The largest of these, called the Monster Group, is roughly of size 8 followed by 53 zeros, more than there are atoms in the earth!

Some amazing connections have been found between the sporadic groups and string theory, the theory that tries to unify the fundamental forces of nature. One formulation of string theory, related to the Monster Group, suggests that the universe has 26 dimensions, not just the four classical dimensions of space and time. These connections are still very speculative, but may one day provide yet another powerful demonstration of how mathematical objects studied for their innate interest and beauty turn out to be crucial in modelling the physical world.

KELVIN WAKES ||

A stone dropped in a pond generates waves in a beautiful changing pattern, a rippling ring radiating outwards from a centre that gradually returns to quietude. The expanding ring is called a wave packet. Individual waves travel at different speeds, the long ones going fastest and the shortest ones slowest. The overall amplitude is greatest where the waves interact with each other constructively to produce the peak of the packet. The speed of this peak is the *group velocity*. We can see the individual waves passing through this peak.

The idea of group velocity was proposed by the Irish mathematician William Rowan Hamilton in 1839 and later analysed more completely by Lord Rayleigh in his *Theory of Sound*. Waves whose speed varies with their wavelength, such as the ripples on a pond, are called dispersive. Fortunately for concert-goers, sound waves are non-dispersive, otherwise a symphony would sound very different at the front and the back of an auditorium.

A stroll in St Stephen's Green provides us with another consequence of wave dispersion, the beautiful wake pattern that follows a swimming duck. The waves spread out in a triangle, with maximum amplitude along the edges of the V-shaped region and a variety of ripples within it. Remarkably, the angle at the apex of the V is fixed at about 40°, irrespective of the duck's speed.

A V-shaped pattern of wake lines is also found behind a boat crossing calm water. Again, the angle between the lines is surprisingly insensitive to the speed of the boat over a wide range of values. This pattern is called a *Kelvin wake*, after Belfast-born William Thomson, later Lord Kelvin. Kelvin knew that in shallow water the wave velocity is the same for all wavelengths. If the boat is travelling slowly, the waves radiate away without any interaction. But if it is moving at a speed faster than the waves, two lines or shock waves form where the waves reinforce each other by constructive interaction. The angle between the shock waves depends on the speed of the boat.

In deep water, things are very different, and shock waves form even when the boat is moving slowly. The long waves move faster than the boat, but the short ones move more slowly. The group velocity is half the wave velocity. Kelvin showed that constructive interference between the waves is concentrated or focused along two lines emanating from the boat, with the angle A between them given by $\sin(A/2) = 1/3$. This means that A is about 40° for the Kelvin wake, whatever the speed of the boat.

Kelvin's analysis went much further, explaining many additional features of the wakes. He accounted for the feather-like pattern of wavelets that form along the wake lines and for the curved transverse waves found within them. Kelvin had many opportunities to study wakes: he was an accomplished sailor, and owned a 126-ton schooner, the *Lalla Rookh*, named after the heroine of a romantic poem by Thomas Moore.

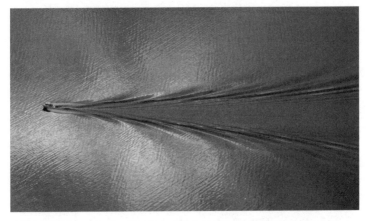

The wake of a boat in the Lyse fjord, Norway. Note that the oblique perspective reduces the apparent size of the wedge angle.

GAUSS MISSES A TRICK

Carl Friedrich Gauss is generally regarded as the greatest mathematician of all time. The profundity and scope of his work is remarkable. So it is amazing that, while he studied non-Euclidian geometry and defined the curvature of surfaces in space, he overlooked a key connection between curvature and geometry. As a consequence, decades passed before a model demonstrating the consistency of hyperbolic geometry emerged.

The Parallel Postulate states that: Given a line L and a point P not on it, there is exactly one line through P parallel to L. Or is it two? Or perhaps none at all? The statement above is John Playfair's version of the fifth postulate of Euclid, the Parallel Postulate.

The Parallel Postulate was a source of frustration and dissatisfaction for centuries. While Euclid's first four postulates are simple, unequivocal and intuitive, the fifth is cumbersome in form and far from self-evident. Concerted efforts were made to prove the parallel property as a theorem based on the other four postulates. But all attempts failed.

NON-EUCLIDEAN GEOMETRY

In the early nineteenth century two mathematicians, one in Hungary and one in Russia, showed that a meaningful version of geometry can be derived in which more than one line through a point P and parallel to L exists. They were János Bolyai and Nikolai Ivanovich Lobachevsky.

Farkas (Wolfgang) Bolyai, father of János, who had studied with Gauss, sent a description of his son's work to the man later described as the Prince of Mathematicians. But Gauss responded that he had obtained similar results decades earlier, although he had not published them. This had a devastating effect on poor János.

Gauss had begun to investigate the Parallel Postulate around 1796 (when he was only nineteen) and had arrived at a system that he later called non-Euclidian geometry. But it is clear that he did not progress as far as Bolyai or Lobachevsky, so his response to Bolyai's father seems rather harsh.

If there was more than one geometry, which was the correct one? Which was a true model of the physical world? Indeed, was the new geometry, which Felix Klein later called hyperbolic geometry, meaningful at all?

Around 1868, the Italian mathematician Eugenio Beltrami showed that hyperbolic geometry could be modelled using a pseudosphere. That is, the axioms and theorems of the new geometry are true for geodesic curves on this surface. The pseudosphere is a surface of constant negative curvature.

Gauss studied the properties of curved surfaces embedded in space, profoundly transforming and extending the subject of differential geometry. He defined the curvature of a surface, relating it to the curvature of the intersections of the surface with planes containing the normal to it.

Gauss discovered an amazing result that he called the Theorema Egregium ('remarkable theorem'). This showed that curvature is intrinsic to the surface: it could be calculated within the surface itself, without reference to the embedding space.

Gauss studied surfaces having constant curvature, positive like the sphere or ellipsoid, zero like the plane or cylinder, and negative like the pseudosphere or hyperbolic paraboloid. But he failed to make the link with non-Euclidian geometry. Had he done so, he could have constructed a model of the hyperbolic geometry of Bolyai and Lobachevsky. Because of this oversight, Gauss never realised that a demonstration of the consistency of the new geometry was there at his fingertips.

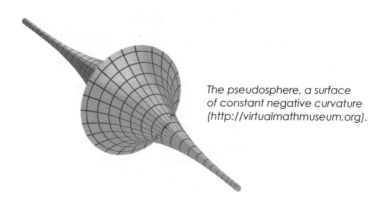

The pseudosphere, a surface of constant negative curvature (http://virtualmathmuseum.org).

SOURCES

Gray, Jeremy (2007), *Worlds Out of Nothing*. New York: Springer.

Sossinsky, A. B. (2012), *Geometries*. Providence, Rhode Island: American Mathematical Society.

PRIME SECRETS REVEALED

Exciting things have been happening in number theory recently. The mathematics we study at school gives the impression that all the big questions have been answered: most of what we learn has been known for centuries, and new developments are nowhere in evidence. In fact, research in maths has never been more intensive and advances are made on a regular basis.

Recently, two major results in prime number theory were announced. Recall that a prime number is one that cannot be broken into smaller factors. So 5 and 7 are prime, but 6 is not, since 6 equals 2 multiplied by 3. Primes are the atoms of the number system. Euclid showed, some 23 centuries ago, that there is an infinitude of primes, but many fundamental questions about their properties remain unanswered. For example, prime pairs like 17 and 19, differing by 2, may be finite or unlimited in number. No one has the answer to this 'Twin Prime' problem.

Another puzzle surrounds the splitting of even numbers into sums of primes, like 10 = 3 + 7. Is this possible for every even number? Christian Goldbach thought so and said as much in 1742 in a letter to his friend Leonhard Euler, the great Swiss mathematician. Euler responded that he regarded Goldbach's conjecture as 'virtually certain, though I cannot prove it'. And no one has proved it since.

Why should we worry? Number theory is the purest of pure mathematics, remote from our daily cares. How could it have any relevance to practical life? In fact, the properties of prime numbers underlie all modern cryptography, which is vital for the integrity of online communications and the security of internet financial transactions.

In 2013, Yitang Zhang of the University of New Hampshire sent a paper to the pre-eminent journal *Annals of Mathematics*, claiming that there are an infinite number of prime pairs whose separation is less than a fixed constant. The constant is huge, about 70 million, a long way from 2, the value needed to prove the twin prime problem. Still, it is a dramatic breakthrough, and ways will soon be found to reduce the separation constant. Zhang's paper was fast-tracked for review and within three weeks one referee had described it as 'first-rate'.

On that very same day, Harald Helfgott of École Normale Supérieure in Paris posted a 133-page pre-print proving a weak form of Goldbach's conjecture: every odd number from 7 upwards is the sum of three primes. This would follow from Goldbach's statement about even numbers, but the argument does not work the other way round, so the original conjecture remains open.

There is cast-iron evidence that Goldbach was right: all even numbers up to 4×10^{18} (four million million million) have been shown to be sums of two primes, but the greatest mathematicians have been unable to prove that the assertion is true in all cases. So the matter remains open; perhaps you can win enduring fame by proving this 275-year-old conjecture.

AMAZING NORMAL NUMBERS ‖

For any randomly chosen decimal number, we might expect that all the digits, 0, 1, ..., 9, occur with equal frequency. Likewise, digit pairs such as 21 or 59 or 83 should all be equally likely to crop up. Similarly for triplets of digits. Indeed, the probability of finding any finite string of digits should depend only on its length. And, sooner or later, we should find any string. This is obviously not true for all numbers; for example, the number 0.11111111 ... with only ones is clearly not normal. But there are compelling arguments that *most* numbers are normal; the word 'most' here has a technical meaning, but we may interpret it in an intuitive way.

To give a precise definition, a decimal number x is *normal* if its digits follow a uniform distribution: that is, all digits are equally likely, all pairs of digits are equally likely, all triplets of digits are equally likely, and so on (strictly, we should say *normal in base 10*). We would expect each of the strings 0, 1, 2, ..., 9 in a normal number to occur one-tenth of the time, each of the strings 00, 01, ..., 98, 99 to occur one-hundredth of the time, and so on.

It is difficult to prove whether a given number is or is not normal. For example, we don't know if the mathematical constants π and e are normal or not. But normal numbers can be constructed. One of the first shown to be normal was the *Champernowne constant*, devised in 1933 by the English economist and mathematician David Champernowne while he was an undergraduate at Cambridge. It is written simply as

$$C = 0 . 1 2 3 4 5 6 7 8 9 10 11 12 13 14 15 16 \ldots$$

where we write all the counting numbers in order after the decimal point (the spaces are for clarity). It is obvious that any string of digits must occur in C, since every string is also a whole number. In fact, every string must occur an unlimited number of times.

Now consider the word IRELAND. We can encode this by assigning numbers to the letters of the alphabet, A=01, B=02, ... , Z=26, yielding the string [09 18 05 12 01 14 04]. This string occurs in C, so we can say that the word IRELAND is encoded in the number. But a similar argument holds for a longer string, such as the Bible or the complete works of Shakespeare. The (very long) string corresponding to the Bible occurs infinitely often in C. And also strings corresponding to translations of the Good Book into every known language. And every other book that has ever been published is hidden within it.

But there is more: your genetic make-up is determined by the structure of your DNA, and this can be expressed as a string of digits. So you are encoded in C, and so am I, and so is everyone else who has ever lived. Indeed, if the universe is finite, it can be represented by a number, unimaginably large but finite nonetheless. So the total state of the universe is in C. One more thing: the winning numbers of next week's lottery are most certainly contained (infinitely often) in C. But there is a slight difficulty: how to get them out!

HEAVY METAL OR BLUE JEANS?

Every June the results of the final examinations in mathematics are read out at the Senate House in Cambridge University. Following tradition, the class list is then tossed over the balcony, and the name of this year's Wranglers will be known. The Wranglers could be a rock group or a brand of American jeans, but they are also the students who gain first-class honours degrees in the examinations known as the Mathematical Tripos, the student ranked first being Senior Wrangler.

Great prestige attaches to the top few Wranglers, opening opportunities for their future professional careers. To become Senior Wrangler was once regarded as 'the greatest intellectual achievement attainable in Britain'. In the past, the rankings in the exams were made public. Since 1910, only the class of degree has been given, but the examiner tips his hat when announcing the name of the top student.

The notoriously difficult Tripos was a test of speed and well-practised problem-solving techniques, and many brilliant students who were inadequately drilled failed to top the class. To have any hope, students needed to be coached like racing thoroughbreds. The 'Old Tripos' tested the mettle of the strongest students. In 1854, when James Clerk Maxwell was beaten into second place by Edward Routh, the Tripos comprised 16 papers over eight days – more than 40 hours in total. Routh went on to become the most successful coach, training 27 Senior Wranglers. Maxwell made monumental contributions to the theory of electromagnetism.

During the nineteenth century, mathematics in Britain lagged behind developments in Germany and France. One of the most inventive and original students who did not make Senior Wrangler was G. H. Hardy, the leading British mathematician of the twentieth century. Hardy placed some blame for Britain's poor performance on the Mathematical Tripos, stating that it was poor training for a pure mathematician. Substantial reforms, introduced in 1909, changed the nature of the Tripos. Prior to that, applications such as Maxwell's electromagnetic theory had been emphasised. Hardy was largely responsible for the change of focus to more pure mathematics.

There were two individuals who were ranked number one but who did *not* become Senior Wrangler. One was Philippa Fawcett, who in 1890 was declared to be 'above the Senior Wrangler'. Her marks were 13% ahead of the next in line but, while women were permitted to take the examinations, they were not allowed at that time to be members of the university or to receive degrees. The other was the Hungarian-born mathematician George Pólya.

He had contributed to the reform of the Tripos and, at the request of Hardy, had sat the examination in 1925. To the great surprise of Hardy, Pólya achieved the highest mark, which, were he a student, would have made him Senior Wrangler.

George Stokes and William Thomson, two Irish-born scientists, were both Wranglers. Stokes was Senior Wrangler in 1841. Thomson, later Lord Kelvin, reckoned himself a 'shoo-in' for the honour in 1845. According to legend, he dispatched one of the college servants thus: 'Just pop down to the Senate House and see who is Second Wrangler.' The servant returned with the answer: 'You, Sir!'

THE SCHOOL OF ATHENS ‖

The word 'geometry' means 'earth measurement', and the subject evolved from measuring plots of land accurately and also from the work of builders and carpenters. So the geometry that we call Euclidean emerged from the needs of artisans. Another form of geometry – projective geometry – was inspired by artists who wished to represent things not as they are but as they look.

We all know that a circular coin appears oval in shape when viewed from an angle. The Greeks were aware of such distortions. Nothing remains of their drawings and paintings, less durable than their sculpture. We know, however, from literary references, that they understood the laws of perspective and used them in designing realistic scenery for their plays.

Western artists rediscovered perspective during the early Renaissance. Piero della Francesca wrote on the use of vanishing points to depict depth, and Filippo Brunelleschi, who designed the magnificent dome of the cathedral in Florence, gave artists the mathematical means of realistically representing three dimensions in painting. The laws of perspective were systematised by Leon Battista Alberti, who dedicated his work 'On Painting' to Brunelleschi.

Raphael's masterpiece *The School of Athens* is of particular interest. This fresco, in the Apostolic Palace in the Vatican, shows all the major classical Greek philosophers and mathematicians, engaging in dialogue or immersed in contemplation. The picture gives a brilliant sense of space through the use of perspective. The setting is an imposing hall with lofty arches, ornate ceiling and mosaic floor, all rendered in proper spatial relationship. There are some forty people in the picture. Plato and Aristotle stand centre-stage, right at the vanishing point of the architectural backdrop. The main figures at left and right foreground are assumed to be Pythagoras and Euclid. The picture, painted around 1510, is a splendid exemplar of the High Renaissance.

The discovery of perspective led to a new form of geometry, called projective geometry. Under the process of projection, distances and angles are distorted. Parallel lines become intersecting lines, just as railway tracks viewed from a bridge converge towards the horizon. Certain properties, however, remain unchanged: a point is projected to a point, a line to a line, and a tangent to a tangent. In this new geometry, we concentrate on the properties that remain unaltered, or invariant, under projection.

From the time of the Impressionists, artists have sought to depict the essence of their subjects, rather than to produce exact likenesses. The cubist painters of the early twentieth century were driven by an urge to paint what they knew to be there rather than what they could directly see. Thus, they developed a variety of techniques to represent three-dimensional forms from multiple viewpoints, unconstrained by the laws of perspective. For the past century, the role of perspective in fine art has been greatly diminished.

Given the visual character of projective geometry, it is no surprise that it emerged from the interests of artists. Today, it is proving vital in another context: computer visualisation. When graphic artists develop computer games, they use projective geometry to achieve realistic three-dimensional images on a screen. And in robotics, automatons use it to reconstruct their environment from flat camera images. Not for the first time, mathematics developed in one context is proving invaluable in another.

‖ HAILSTONE NUMBERS

Hailstones, in the process of formation, make repeated excursions up and down within a cumulonimbus cloud until finally they fall to the ground. We look at sequences of numbers that oscillate in a similarly erratic manner until they finally reach the value 1. They are called hailstone numbers.

There are many simply stated properties of the familiar counting numbers that appear certain to be true but that have never been proved. A prime example is Goldbach's conjecture: every even number greater than 2 is the sum of two prime numbers. In this category, we consider a conjecture made by Lothar Collatz, a German mathematician, in 1937. It is easily stated, but no one has ever proved it.

To explain the conjecture, we construct a sequence of numbers by a simple iterative process. Let the first value be any positive whole number N. If N is even, divide it by two. If it is odd, multiply by 3 and add 1. Thus, we get a new number, either $N/2$ or $3N+1$. The process is often abbreviated HOTPO, for 'Half Or Triple Plus One'.

We can represent this process in terms of a map:

$$n \text{ goes to } n/2, \text{ if } n \text{ is even}$$

$$n \text{ goes to } 3*n+1, \text{ if } n \text{ is odd}$$

Now we repeat the process over and over to generate a sequence. The Collatz conjecture is that, no matter what value N we choose to start, the sequence always reaches 1 after a finite number of steps.

Let's try a few examples: $N = 3$ goes to 10, then 5, then 16, 8, 4, 2 and 1. After this it cycles for ever between 4, 2 and 1, so we stop at the first occurrence of 1, giving a sequence of eight numbers (3, 10, 5, 16, 8, 4, 2, 1). Next, starting with $N = 7$, we reach 1 after 17 steps (7, 22, 11, 34, 17, 52, 26, 13, 40, 20, 10, 5, 16, 8, 4, 2, 1).

So far it has been easy to compute the sequence mentally. But if we choose $N = 27$ there are 112 numbers in the sequence, which reaches 9232 after 78 steps before finally descending to 1. The sequence, shown on the next page, is seen to oscillate wildly, illustrating the reason for the name hailstone numbers.

The Collatz conjecture has been checked for all starting values up to 5×10^{18}, and all cases end in the cycle 4, 2, 1. Of course, this is not proof that the conjecture holds for all N, but it is powerful evidence. It is difficult to doubt that Collatz was right.

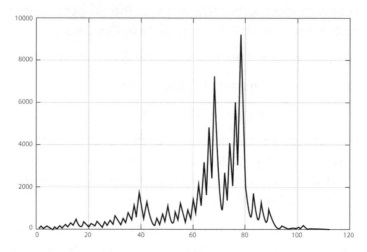

The hailstone sequence for the initial value N = 27. The sequence reaches its maximum value of 9232 at step 78 and arrives at 1 in 112 steps.

THE REMARKABLE BBP FORMULA ‖

Information that is declared to be for ever inaccessible is sometimes revealed within a short period. Until recently, it seemed impossible that we would ever know the value of the quintillionth decimal digit of π. But a remarkable formula has been found that allows the computation of binary digits starting from an arbitrary position without the need to compute earlier digits. This is known as the BBP formula.

EARLY ESTIMATES OF π

For millennia, mathematicians have been intrigued by π. Despite prolonged and intensive study, many simple questions about this number remain unanswered. The digits of π appear to occur with random frequency, but there is no proof of this. We don't even know whether any specific digit, say 4, in the decimal expansion occurs infinitely often.

Archimedes approximated π using polygons drawn within and around a circle. In the plague year of 1666, Isaac Newton evaluated 15 decimal digits of π, remarking apologetically that he 'had no other business at the time'. In 1761 the Swiss mathematician Johann Lambert showed that π is irrational, so its digits do not repeat in any number base. Over a century later, Ferdinand von Lindemann proved that π is transcendental, implying that no polynominal with integer coefficients has a root equal to π. This also demonstrated the impossibility of solving the ancient Greek problem, to construct a square with the area of a given circle.

THE COMPUTER ERA

Since the dawn of the computer era, the number of known digits of π has grown exponentially. In 1949 over two thousand digits of π were computed using ENIAC (the Electronic Numerical Integrator and Computer), and John von Neumann sought patterns in the digits but found none. By 1973 π was known to a million digits, by 1989 to a billion and by 2002 to a trillion. Currently, more than five trillion digits are known.

It is difficult even for experts to anticipate advances in computing techniques and in the efficiency of algorithms for computing π. In his book *The Emperor's New Mind*, Roger Penrose wrote that we would *never* know if a string of ten consecutive 7s occurs in π. But such a string was found within a decade (at about 23 billion digits into the expansion).

THE BBP FORMULA

In 1997 Bailey, Borwein and Plouffe published a remarkable formula for π:

$$\pi = \sum_{k=0}^{\infty} \left[\frac{1}{16^k} \left(\frac{4}{8k+1} - \frac{2}{8k+4} - \frac{1}{8k+5} - \frac{1}{8k+6} \right) \right]$$

Because of the factor 16^k, this allows the direct calculation of the hexadecimal digits of π, beginning at an arbitrary position without any need to compute earlier digits. The discovery of the formula required considerable ingenuity but, once found, its proof is surprisingly simple. For a description of the hunt for the formula and an outline of the proof, see Bailey *et al.* 2013.

The BBP formula came as a surprise. It had not been expected that it would be possible to extract digits far into the expansion without crunching out all the earlier digits. Of course, the formula provides only the hexadecimal digits. Subsequent searches for an analogous formula giving the decimal digits have been unsuccessful and it is now doubted if any such formula exists.

NORMAL NUMBERS

A number is normal if its digits are distributed 'randomly' or, more specifically, if every string of m digits occurs with limiting frequency 10^{-m}. Thus, each digit from 0 to 9 occurs (ultimately) 10% of the time, each pair from 00 to 99 occurs 1%, and so on. A more general definition of normality includes expansions in all number bases.

Proving normality is very difficult except for some specially constructed numbers, for example Champernowne's Number (see page 86). It is not known if π is normal but the numerical evidence points towards this. Recent progress gives us hope that an answer may be found within a decade or two.

SOURCES

Bailey, David, Peter Borwein and Simon Plouffe (1997), 'On the rapid computation of various polylogarithmic constants', *Mathematics of Computation*, **66**, 903–13.

Bailey, David H., Jonathan M. Borwein, Andrew Mattingly and Glenn Wightwick (2013), 'The computation of previously inaccessible digits of π^2 and Catalan's constant', *Notices of the American Mathematical Society*, **60**, 844–54.

Wikipedia, 'Bailey–Borwein–Plouffe formula', https://en.wikipedia.org/wiki/Bailey%E2%80%93Borwein%E2%80%93Plouffe_formula

THE ATMOSPHERIC RAILWAY

For more than ten years from 1843 a train without a locomotive plied the two-mile route between Kingstown (now Dun Laoghaire) and Dalkey. Trains running every 30 minutes were propelled up the 1 in 110 gradient to Dalkey by atmospheric pressure. Returning trains coasted down to Kingstown under gravity.

A fifteen-inch cast iron pipe was laid between the railway tracks and a piston in the pipe was linked through a slot to the front of the train. The slot was sealed by a greased leather flap to ensure that it was airtight. The air in the section of pipe ahead of the train was exhausted by a steam-driven air pump in an engine house at the Dalkey terminus. With a partial vacuum ahead, the atmospheric pressure behind the piston drove the train forward.

Available technical data are scant, but adequate for some back-of-the-envelope calculations to estimate average speeds and travel times. Pressure is force per unit area. The atmosphere presses on the piston from both ends; the net force is the pressure difference multiplied by the area. The area of the piston face follows from its diameter and, with a working vacuum of about 15 inches of mercury, or half an atmosphere, the pressure difference is 500 hPa. Then the force on the piston comes to about 57,000 newtons.

Using Newton's Law, we can divide the force by the mass of the train to get the acceleration. Assuming a train of mass 10 tonnes, the acceleration comes to about 0.6 metric units. This may be compared to the 10 metric units of the acceleration due to gravity but, with the small gradient, only about one per cent of the gravitational acceleration acts along the pipe, so the net acceleration is about 0.5 metric units.

Now a standard equation of elementary mechanics comes to the fore: the square of the terminal speed is twice the acceleration times the distance. Knowing the acceleration and the length of the line (2,800 metres) we can estimate the maximum speed and journey time. The formula gives a maximum speed of 190 km/h and an average speed of half this value. However, we have neglected friction, which reduces this speed considerably. Typical running speeds of 40 km/h would give a journey time for the up train of about four minutes. This is consistent with reported times.

We can also estimate the maximum speed on the downhill journey by invoking energy conservation. At Dalkey, the train has gained potential energy (mass times gravitational acceleration times height). The terminus at the Dalkey end was 25 metres above sea level. Supposing the potential energy were to be converted completely to kinetic energy (half mass times speed squared), the maximum speed would be about 80 km/h. A mean speed of 40 km/h is an overestimate because frictional losses have again been ignored. According to contemporary reports, the return journey under gravity was 'very lady-like', the average speed being about 30 km/h and the journey taking between five and six minutes.

Although reasonably successful, the atmospheric system had several operational inconveniences and was abandoned after about ten years. But a system using similar pneumatic principles is running today in Jakarta and another, called Aeromovel, is operating in Brazil. So air power, which seemed like a white elephant 170 years ago, may again provide fast, clean and frequent urban transport for Dublin.

A HOLE THROUGH THE EARTH

'I wonder if I shall fall right through the earth', thought Alice as she fell down the rabbit hole, 'and come out in the antipathies.' In addition to the author of the Alice books, Lewis Carroll – in real life the mathematician Charles L. Dodgson – many famous thinkers have asked what would happen if one fell down a hole right through the earth's centre.

Galileo gave the answer to this question: an object dropped down a hole piercing the earth diametrically would fall with increasing speed until the centre, where it would be moving at about 8 km per second, after which it would slow down until reaching the other end, where it would fall back again, oscillating repeatedly between the two ends.

Of course, drastic simplifications are being made here, with no account taken of air resistance or the effects of the earth's rotation, not forgetting the engineering challenge of constructing the hole through molten magma under extreme pressures and temperatures. But let us proceed with this fantasy or, as Einstein might have called it, *Gedankenexperiment*.

SIMPLE HARMONIC MOTION

The force due to gravity on an object of mass m at the earth's surface is

$$F = (GMm/a^2) = mg$$

where M and a are the mass and radius of the earth and G is the universal gravitational constant ($G = 6.67 \times 10^{-11} \text{m}^3\text{kg}^{-1}\text{s}^{-2}$), and the acceleration due to gravity is $g = 9.8$ m/s^{-1}. When the object is beneath the surface, at a distance r from the centre, there are two contributions, one from the sphere of radius r beneath the object and one from the spherical shell of thickness $(a - r)$ outside it. For uniform density, the attraction of the inner sphere is as if all its mass were concentrated at the centre. The attraction of the outer shell vanishes; this follows from a simple symmetry argument showing that the attractions of opposite sides of the shell cancel. So the force at radius r is

$$F(r) = (GM(r)m/r^2) = (mg/a)\, r$$

where $M(r)$ is the mass of the inner sphere of radius r. The equation of motion for the object is now

$$d^2r/dt^2 + (g/a)\, r = 0$$

which is the equation for simple harmonic motion with frequency $\omega = \sqrt{(g/a)}$. The half-period, which is the time to travel through the tunnel, is

$$\tau/2 = \pi \sqrt{(a/g)}$$

which, with $a = 2 \times 10^7/\pi$ m, comes to 42.2 minutes. This is a very remarkable result. It also follows that the maximum speed, reached at the earth's centre, is $a\omega = \sqrt{(ag)}$ with a value of 7900 m/s or about 8 km per second. If you thought it was hot there, it's just got hotter!

Suppose now that the tunnel runs straight between two points not diametrically opposite each other, so not through the earth's centre. The component of gravity along the route is now weaker, making for a longer trip. But the distance is less, making the time shorter. These two effects cancel exactly, giving us the amazing result that the time to traverse any straight tunnel through the earth is 42.2 minutes, regardless of the tunnel's length.

Two curiosities about the motion of an object oscillating in a hole through the earth:

- the period depends only on the density of the planet, not on its radius or total mass.

- the period is precisely that of a satellite orbiting the earth at ground level.

RAPID RAIL TRANSPORT

In another of Lewis Carroll's books, *Sylvie and Bruno Concluded*, the author describes an ingenious method of running trains, based on the above ideas, with gravity as the driving force. The track would be laid in a straight tunnel between two cities. Trains would run downhill to the midpoint, which is closer than the ends to the earth's centre, gaining enough momentum to carry them uphill to the other end. It is a simple but surprising mechanical result that the journey time is independent of the distance. Two trains leaving Dublin at midday, one bound for London and one for Sydney, would both arrive at 12:42, less than three-quarters of an hour later.

SOFIA KOVALEVSKAYA ‖

In the nineteenth century it was extremely difficult for a woman to achieve distinction in the academic sphere, and virtually impossible in the field of mathematics. But a few brilliant women managed to overcome all the obstacles and prejudice and reach the very top. The most notable of these was a remarkable Russian, Sofia (also known as Sonya) Kovalevskaya (née Korvin-Krukovskaya).

Sofia was born in Moscow in 1850, into an aristocratic and intellectually gifted family. When their house at Palibino was redecorated, there was insufficient wallpaper, and the nursery walls were covered with lithographed lecture notes on calculus that her father had retained from his student days. Sofia recalled reading these 'mathematical hieroglyphics' with interest and wonderment, and they provided inspiration and impetus for her later career.

There was no opportunity for academic advancement in Russia for Sofia. Moreover, it was difficult for a single woman to travel unescorted. So, when she was just eighteen, a marriage of convenience was arranged with a young palaeontologist, Vladimir Kovalevsky. The following year, the couple travelled to Heidelberg, where Sofia obtained permission to attend lectures at the university.

In 1870 Sofia moved to Berlin. Regulations there prohibited her from auditing lectures but she received private tuition from the great mathematician Karl Weierstrass. He was enormously impressed by her dazzling intelligence. The professor, who was fifty-five and unmarried, was also smitten by the luminescent vivacity of the twenty-year-old Sofia, and they developed a warm personal relationship.

Under the supervision of Weierstrass, Sofia began research on partial differential equations and in 1874 she was awarded a doctoral degree *summa cum laude* by Göttingen University. Her work, which extended that of the Frenchman Augustin-Louis Cauchy, is known today as the Cauchy–Kovalevskaya Theorem.

Sofia and Vladimir agreed that their nominal marriage could become a genuine one and the couple returned to St Petersburg, where they had a daughter in 1878. But opportunities in Russia were few and, when Vladimir's business affairs went awry, he took his own life.

At that time, European universities were at last beginning to open their doors to women. Sofia obtained a temporary lectureship in Stockholm, where she earned a reputation as a superlative teacher. She won the prestigious Prix Bordin of the French Academy of Science for her research. The problem she solved is described in advanced mechanics texts as the Kovalevskaya top, an asymmetric gyroscope, one of the very few problems in dynamics that can be completely solved. The judges were so impressed that they increased the prize from 3000 to 5000 francs.

Sofia's brilliant research secured her a professorship in Stockholm. She was elected a corresponding member of the Russian Academy of Sciences, smashing another gender barrier. She also became an editor of *Acta Mathematica*, the first woman on the board of a scientific journal. But her career was to be cut short: already shocked by her husband's suicide, she was shattered by the early and sudden death of her sister Aniuta. Sofia became ill with pneumonia and died, aged just forty-one, in 1891.

Sofia Kovalevskaya's mathematical work is of enduring value. She was the first woman to be awarded a PhD in mathematics and the first female professor of mathematics. She was a leader in the movement for the emancipation of women and did much to improve women's access to the academic arena.

THE SIMPLER
THE BETTER

A Berkeley graduate student, George Dantzig, was late for class. He scribbled down two problems written on the blackboard and handed in solutions a few days later. But the problems on the board were not homework assignments; they were two famous unsolved problems in statistics. The solutions earned Dantzig his PhD.

With his doctorate in his pocket, Dantzig went to work with the US Air Force designing schedules for training, stock distribution and troop deployment, activities known as programming. He was so efficient that, after World War II, he was given a well-paid job at the Pentagon with the task of mechanising the military's programme planning. There he devised a dramatically successful technique, or algorithm, which he named *linear programming* (LP).

LP is a method for decision-making in a wide range of economic fields. Industrial activities are often limited by constraints. For example, there are normally constraints on raw materials and on the number of staff available. Dantzig assumed these constraints to be linear, with the variables, or unknown quantities, occurring in a simple form. This makes sense: if you need four tons of raw material to make 1000 widgets, then eight tons will be needed to make 2000. Double the output requires double the resources.

LP finds the maximum value of a quantity such as output volume or total profit, subject to the constraints. This quantity, called the objective, is also linear in the variables. A real-life problem may have hundreds of thousands of variables and constraints, so a systematic method is needed to find an optimal solution. Dantzig devised a method ideally suited to LP, called the *simplex method*.

At a conference in Wisconsin in 1948, when Dantzig presented his algorithm, a senior academic objected: 'But we all know the world is nonlinear.' Dantzig was nonplussed by this put-down, but an audience member rose to his defence: 'The speaker titled his talk "Linear Programming" and carefully stated his axioms. If you have an application that satisfies the axioms, then use it. If it does not, then don't.' This respondent was none other than John von Neumann, the leading applied mathematician of the twentieth century.

The acid test of an algorithm is its capacity to solve the problems for which it was devised. LP is an amazing way of combining a large number of simple rules and obtaining an optimal result. It is used in manufacturing, mining, airline scheduling, power generation and food production, maximising efficiency and saving enormous amounts of natural resources every day. It is one of the great success stories of applied mathematics.

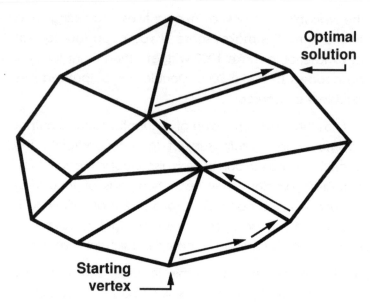

The simplex algorithm moves along the edges of the polytope until it reaches the optimum solution.

GEOMETRY OUT OF THIS WORLD ‖

The shortest distance between two points is a straight line. This is one of the basic principles of Euclidean geometry. But we live on a spherical earth, and we cannot travel the straight-line path from Dublin to New York: we have to stick to the surface of the globe, and the geometry we need is more complicated than the plane geometry of Euclid. Spherical geometry is central to the study of geophysics and astronomy, and vital for navigation.

We can define a metric for the distance between two neighbouring points on the globe and use it to find the shortest route between them. Paths of minimum distance are called 'geodesics' and they play the role of straight lines on the earth. They are the great circles with centres at the earth's centre. Flights from Paris to Vancouver pass over Greenland, far to the north of both cities, following a great circle route.

The three angles of a plane Euclidean triangle add to two right angles or 180 degrees. A triangle on the globe has three angles whose sum exceeds two right angles.

Moreover, there are no parallel lines, since two great circles always intersect. The earth is embedded in three-dimensional Euclidean space, so nobody thought of spherical geometry as different, even though this was under our noses, hidden in plain sight. It took the genius of Gauss to spot its fundamental nature and to introduce the term 'non-Euclidean geometry'.

Spherical geometry is just one flavour of non-Euclidean geometry. Defining other metrics for the distance between neighbouring points, we can construct entirely new geometries. If we use the paths of light as our straight lines or geodesics, we get a new form of geometry quite distinct from either spherical geometry or the plane geometry of Euclid.

In physics, we learn that light travels in a straight line, but this is true only in a homogeneous medium. Where there are variations in the medium, such as the changing density of the atmosphere with height, light rays may bend, with interesting consequences. The downward-curving light rays define another form of geometry, called hyperbolic geometry.

Since the earth is spherical, horizontal light rays can travel large distances if their curving path is about the same as the curvature of the earth. In unusual cases, where the air near the surface is very cold and dense, the curving light can come from beyond the horizon. Mariners may see vessels just over the horizon, or more distant land masses. Once or twice in a decade, such weather conditions enable observers in the Dublin mountains to see the peaks of Snowdonia in north Wales.

Hyperbolic geometry has been studied intensively ever since it was formulated independently by three mathematicians, Bolyai, Lobachevsky and Gauss, almost two hundred years ago. Bolyai described his breakthrough in a letter to his father: 'I have created a new and different world out of nothing.'

We no longer doubt the spherical shape of the earth, but the shape of the universe is a hot topic amongst cosmologists. General relativity tells us that matter causes space to curve, resulting in locally non-Euclidean geometry. But the global shape, or topology, of space remains a mystery. Perhaps some entirely new geometry will be required to describe it.

‖ EULER'S GEM

The highlight of the thirteenth and final book of Euclid's *Elements* is the proof that there are just five 'Platonic solids'. Recall that a regular polygon is a plane figure with all sides and angles equal, for example a square. By joining together identical polygons, we can form solid bodies called regular polyhedra.

Thus, with six squares we can make a cube. Four equilateral triangles make a tetrahedron, eight make an octahedron and 20 make an icosahedron. The fifth and final regular polyhedron is the dodecahedron, with 12 pentagonal faces.

Plato used the regular polyhedra in an early atomic model and we still speak of the Platonic solids. Archimedes discovered that by using more than one type of polygon he could form 13 new 'semi-regular' solids. Much later, Kepler constructed a magnificent but incorrect model of the solar system using the Platonic solids.

A simple property of all these shapes escaped the notice of the ancient mathematical luminaries. If we denote the number of vertices, edges and faces by V, E and F respectively, then

$$V - E + F = 2$$

This relationship was found by the Swiss mathematician Leonhard Euler in 1750, and is now called Euler's polyhedron formula or, more colloquially, Euler's Gem.

And a gem it is: although simple enough to explain to a young child, the formula has sweeping implications throughout mathematics, with repercussions for topology, network analysis and dynamical systems theory.

It is easy to check that a cube has eight vertices, 12 edges and six faces, so the formula holds. But it applies much more generally. Any surface can be divided up into a mesh of interlocking triangular pieces; cartographers construct maps using this process of triangulation. Thus, a curved surface is approximated by a polyhedron with flat faces. No matter how the globe is triangulated, Euler's formula holds.

Euler's Gem provides a bridge between geometry and topology. In geometry, lengths and angles are fixed. We can move shapes around but no distortion is permitted. Topology is geometry with rigidity relaxed. Shrinking and stretching are allowed, but no tearing or gluing. So, a sphere and cube are topologically equivalent: a round ball of plasticine can be squashed into cubic shape. But a sphere and torus (doughnut shape) are distinct: we cannot change one into the other without making or filling a hole.

A torus can be triangulated just like a sphere, but now Euler's number $X = V - E + F$ becomes 0 instead of 2. More generally, if we triangulate a surface and find Euler's number X, we can be sure that there are N = (1 – X/2) holes in it.

Euler's formula has some sporting implications. One brand of golf ball has 232 polygonal dimples, all hexagons except for 12 pentagons. And the football popularised in the 1970 World Cup in Mexico was based on a truncated icosahedron with 20 hexagons and 12 pentagons. Since then, chemists have synthesised a molecule of this shape comprising 60 carbon atoms, earning a Nobel Prize and starting a new field of chemistry.

Some computer models used to simulate the earth's climate use hexagonal meshes to achieve a uniform distribution throughout the globe, circumventing the problem of polar clustering with conventional latitude–longitude grids. But hexagons alone will not suffice. Euler's formula constrains all these grids to have precisely 12 pentagons.

Football (right) based on the truncated icosahedron.

THE WATERMELON PUZZLE ‖

An amusing puzzle appears in a 2013 book, *X and the City*, by John A. Adam. A farmer brings a load of watermelons to the market. Before he sets out, he measures the total weight and the percentage water content. He finds that the total weight is 100 kg and the water content is 99%.

The weather is hot, so his load loses some moisture en route. He checks the water content when he arrives at the market: it has dropped to 98%.

Question: *What is the total weight of the load on arrival at the market?*

Answer: Initially there is **99%** water and so **1%** pith. At the market, there is **98%** water, and therefore **2%** pith. *The percentage of pith has doubled.* But the actual amount of pith is unchanged. The only way this can happen is if the *total weight is halved*. So the answer is **50 kg**. Surprised? The answer is quite surprising. Most people would guess something around 98 or 99 kg. The illustration on the next page should make everything clear:

The upper panel shows the initial weight. The total weight is 100 kg, comprising 1 kg pith (black) and 99 kg water (grey). The lower panel, upon arrival at market, shows the total weight of 50 kg, comprising 1 kg pith (black) and 49 kg water (grey). In terms of percentages, the upper panel, with 99 kg of water, makes up 99% of the total. The 1 kg of pith makes up just 1%. In the lower panel, 49 kg water makes up 98%, and 1 kg of pith makes up 2% of the load.

The weight of pith does not change, but the percentage doubles.

SOURCE

Adam, John A. (2013), *X and the City: Modeling Aspects of Urban Life*. Princeton, NJ: Princeton University Press.

THE ANTIKYTHERA MECHANISM: THE FIRST COMPUTER ‖

Two storms, separated by two thousand years, resulted in the loss and the recovery of one of the most amazing mechanical devices made in the ancient world. The first storm, around 65 BC, wrecked a Roman vessel bringing home booty from Asia Minor. The ship went down near the island of Antikythera, between the Greek mainland and Crete.

The second storm, in 1900, forced some sponge divers to shelter near the island, where they discovered the wreck. This led to the first ever major underwater archaeological expedition. In addition to sculptures and other artworks, an amorphous lump of bronze, later described as the Antikythera Mechanism (AM), was found.

On examination, the bronze lump turned out to be a complex assemblage of gears, a mechanical device previously unknown in Greek civilisation. Inscribed signs of the Zodiac suggested that it was probably for astronomical rather than navigational purposes.

Several techniques were used to establish that the AM is about two thousand years old. Carbon dating of the ship's timber put its origins at around 200 BC, but the wreck could have happened many decades later. The style of amphora jars found on board implied a date between 86 and 60 BC. Coins found in the wreckage allowed this to be pinned down to about 65 BC.

The inscriptions on the AM link it to Corinth and thence to its colony at Syracuse, where Archimedes flourished. This gives an intriguing possibility that the AM was in a mechanical tradition inspired by Archimedes.

The AM was driven by a handle that turned a linked system of more than 30 gear wheels. Using modern imaging techniques, it is possible to count the teeth on the wheels, see which cog meshes with which and what are the gear ratios. These ratios enable us to figure out what the AM was computing.

The gears were coupled to pointers on the front and back of the AM, showing the positions of the sun, moon and planets as they moved through the zodiac. An extendable arm with a pin followed a spiral groove, like a record player stylus. A small sphere, half white and half black, indicated the phase of the moon.

Even more impressive was the prediction of solar and lunar eclipses. It was known to the Babylonians that if a lunar eclipse is observed, a similar event occurs 223 full moons later. This period of about 19 years is known as the Saros cycle. It required complex mathematical reasoning and advanced technology to implement the cycle in the AM.

The AM could provide accurate predictions of eclipses several decades ahead. Derek de Solla Price, who analysed it in the 1960s, said the discovery was like finding an internal combustion engine in Tutankhamen's tomb.

The Antikythera mechanism has revolutionised our thinking about the scientific legacy of the Greeks. It is like modern clockwork, but clocks were invented in medieval Europe! It shows that the Greeks came close to our technology. Had the Romans not taken charge, we might today be far in advance of our current level of technology.

All the gear ratios are now understood; there was even a dial to indicate which of the pan-Hellenic games would take place each year, with the Olympics occurring every fourth year. Just one small cog remains a mystery. Research is ongoing and more remains to be discovered about this amazing high-tech device.

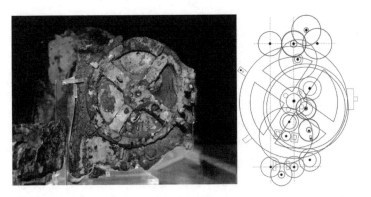

Left: Fragment A of the Antikythera Mechanism. Right: Diagram of the gearing of the AM.

WORLD POPULATION

The Population Division of the United Nations marked 31 October 2011 as the 'Day of Seven Billion'. While that was a publicity gambit, world population is now above this figure and climbing. The global ecosystem is seriously stressed, and climate change is greatly aggravated by the expanding population. Accurate estimates of growth are essential for assessing our future well-being.

Before the Black Death in the fourteenth century, there were about 450 million people. After the Plague this figure had fallen to 370 million. Since then there has been continuous growth. In percentage terms, the growth rate peaked at 2.2% per annum in 1963. It has now declined to about 1.1%. Population growth varies widely from place to place but in most countries there has been a demographic transition to smaller families in recent decades.

The UN estimated a growth rate of 75 million per annum in 2000, and the CIA's *World Factbook* gave a comparable figure. Global life expectancy is about 67 years and rising. Death rates change dramatically due to disease, war and natural disasters. Better sanitation and medical advances reduce infant mortality and more efficient agricultural practice reduces the incidence of famine, so life expectancy is on the increase.

Many mathematical models show an exponential growth; this implies a fixed time for the population to double. In 1798 Thomas Malthus predicted that the exponentially growing population would exhaust the world food supply within fifty years. But, although scary, this may not be pessimistic enough: during the second millennium, each doubling of population took half as long as the previous one. This pattern corresponds to 'hyperbolic growth', with the population becoming unbounded within a finite period, which is much more worrying than exponential growth.

At the more optimistic end of the spectrum we have 'logistical growth', where there are constraints on the maximum population. An early model for this was proposed by the Belgian demographer Pierre François Verhulst, who refined the model of Malthus, introducing a number called the *carrying capacity*. This is the maximum population that the environment can sustain. The resulting growth is initially exponential, but flattens out as it approaches the carrying capacity. Alas, there is little agreement about the earth's carrying capacity.

What lies ahead? A balanced plateau or a Malthusian catastrophe with mass starvation? Current projections all indicate an increasing population, but forecasts vary widely since they depend sensitively on the underlying statistical assumptions. There is considerable uncertainty, so, in addition to actual numbers, probabilistic predictions provide a measure of confidence in the forecasts. The best current mathematical models give expected values for 2100 ranging from 9 billion to 13 billion, with the most probable value being 11 billion.

Total fertility rate is a critical factor in determining growth, and this has been falling in recent decades. As a result, experts consider it unlikely that there will be another doubling of population in the twenty-first century. The population is now seven times larger than when Malthus uttered his gloomy and inaccurate prognosis. Food production has increased dramatically, much more rapidly than he anticipated. But this has come at great cost in terms of environmental stress, pollution and global warming. There is enough food to go round; the problem is that it doesn't!

IRELAND'S FRACTAL COAST ||

Reports of the length of Ireland's coastline vary widely. The CIA's *World Factbook* gives a length of 1448 km; the Ordnance Survey of Ireland 3171 km (www.osi.ie); the World Resources Institute, using data from the United States Defense Mapping Agency, 6347 km.

How come the values differ so much? It is because the coastline is fractal in nature and the measured length depends strongly on the 'ruler' or unit of length used. A straight line – like a road – has dimension 1, and a plane surface – like a field – has dimension 2. But some curves – like coastlines – are so wiggly that they have a dimension between 1 and 2. Such curves, with fractional dimensions, are called fractals.

An early example of a fractal curve is the Koch snowflake, which first appeared in a 1904 paper entitled 'On a continuous curve without tangents, constructible from elementary geometry' by the Swedish mathematician Helge von Koch. It is built up by repeatedly subdividing the sides of an equilateral triangle into three parts and erecting a smaller triangle on the central segment. Here are the first four stages:

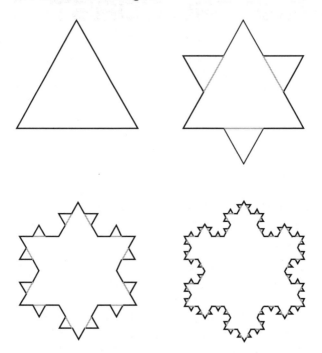

The length of the Koch snowflake grows without limit as the subdivision process continues: each step adds one-third, so at stage N the length is $(4/3)^N$ times the length of the initial triangle, growing exponentially with N. The dimension of the snowflake is log 4 / log 3 ≈1.26.

Fractals arise when we try to assign a length to a highly indented coastline such as the coast of Norway. Larger-scale maps include finer detail, resulting in a longer coastal length. Since we could choose to include variations corresponding to every rock, pebble or grain of sand, it is impossible to assign a length unequivocally.

All we can do is to describe how the length varies with our 'ruler' or unit of measure.

Lewis Fry Richardson measured the coastal length by walking dividers with fixed distance Δ along a map. He found that the dependence of measured length L on the distance Δ was

$$L = C \times \Delta^{(1-D)}$$

For a very smooth coast like that of South Africa, $D \approx 1$. For more convoluted coastlines, $D > 1$, and so L increases as Δ decreases. Richardson found that the west coast of Britain yielded a value $D = 1.25$, close to the dimension of the snowflake curve.

A group of students in the School of Physics at Trinity College Dublin, under the supervision of Professor Stefan Hutzler, studied Ireland's coastline using Google Maps and a measure called the *box dimension*: the length of a curve is estimated by superimposing a square grid on it and counting the number N of grid boxes that contain a segment of the curve. For a smooth curve, the product of N and the width W of a box is not sensitive to the width of the box, and gives the length of the curve. For fractal curves, the length increases as the box width is reduced.

Hutzler obtained a value $D = 1.2$ for the fractal dimension. Not surprisingly, the ragged Atlantic shore has a higher fractal dimension ($D = 1.26$) than the relatively smooth east coast ($D = 1.10$).

Independently, and using a different method, a group in the School of Computing and Mathematics, University of Ulster used printed maps and dividers of various lengths.

This yields what they call the *divider dimension*. They found a value D = 1.23 for the overall coastline. Their length for a divider step of 1 km was 3175 km, very close to the value of 3171 km given by the Ordnance Survey of Ireland (also corresponding to a step size of 1 km).

The close agreement between the two independent studies confirms our ideas about the fractal nature of the Irish coastline and gives us confidence in the reported values of D.

SOURCES

Hutzler, S. (2013), 'Fractal Ireland', *Science Spin*, **58**, 19–20.

McCartney, M., G. Abernethy and L. Gault (2010), 'The divider dimension of the Irish coast', *Irish Geography*, **43**, 277–84.

Wikipedia: 'List of countries by length of coastline', https://en.wikipedia.org/wiki/List_of_countries_by_length_of_coastline

SANTA'S FRACTAL JOURNEY ‖

How far must Santa travel on Christmas Eve? At a broad scale, he visits all the continents. In more detail, he travels to every country. Zooming in, he goes through each city, town and village and ultimately to every home where children are asleep. The closer we examine the route, the longer it seems. This is a characteristic of paths or graphs called fractals.

Smooth geometric curves such as circles and ellipses have dimension 1 and a definite length can be assigned to them. Other curves are so twisted and contorted that their dimension has a fractional value between 1 and 2. They are called fractals.

Fractals were first considered by the English Quaker mathematician Lewis Fry Richardson. He was studying the factors contributing to warfare and thought that the length of the borders between neighbouring countries might be relevant. But when he investigated the frontier between France and Spain, he found that widely different lengths were reported. Similarly, when he measured the west coast of Britain, the length varied with the scale of the map he used.

As discussed in 'Ireland's Fractal Coast' (page 127), larger-scale maps include finer detail, resulting in a longer coastal length. Since there are variations on every scale, it is not possible to assign a definite length in an unambiguous way. Instead, we describe how the length varies with our choice of 'ruler' or unit of measure. This is determined by the fractal dimension. Mathematicians can assign a numerical dimension D to an irregular curve or surface, and it need not be a whole number.

Classical mathematics treated the regular geometric shapes of Euclid and the smoothly evolving dynamics of Newton. Fractal structures were regarded as 'pathological', but, following the inspiration of Benoit Mandelbrot, we now realise that they are inherent in many objects of nature – clouds, ferns, lightning bolts, our blood vessels and lungs, and even the jagged price curves of the stock market.

Fractals are all around us at Christmas; just consider the branching structure of the Christmas tree, or the snowflakes hanging on it. Even the broccoli on your dinner plate is fractal. But what of Santa's route?

If we assume the route is fractal, most of the distance is due to the small segments from one house to the next. With a billion houses to visit, and the typical distance between neighbouring houses being ten metres, we get a length of 10 million km. This is equivalent to about 250 laps of the globe. The long stages, from the North Pole to Kiribati in the central Pacific, Santa's first point of call, and the ocean crossings, contribute very little. It is the small-scale hops from house to house that count for most.

INTERESTING BORES

According to the old adage, water finds its own level. But this is true only for still water. In more dynamic situations, where water is moving rapidly, there can be sudden jumps in the surface level. When water at high speed surges into a zone of slower flow, an abrupt change of depth, called a hydraulic jump, may occur. Hydraulic jumps are found in some river estuaries, below dams and weirs and even in the kitchen sink.

When the tide floods into a funnel-shaped river estuary, it can form a wall of water a metre high, called a tidal bore, that travels rapidly upstream. The Severn bore occurs several times each year, under suitable conditions of spring tide.

Tidal bores are popular with intrepid surfers, who ride them upriver for miles. But there are also stationary hydraulic jumps, found at the spillway of a dam or below a weir. These have recirculating water that traps flotsam and they provide entertainment for kayakers prepared to risk being caught in the back-flow.

A simple mathematical analysis yields information on the height of a hydraulic jump. An early study was carried out by Jean-Baptiste Bélanger (1790–1874), a French engineer, who derived an expression for the difference in height by means of conservation principles. Gustave Eiffel was one of his students, and Bélanger's name is inscribed on the Eiffel Tower.

It might seem reasonable to apply the principle of conservation of energy to the analysis. But energy is dissipated at the hydraulic jump, so the conservation principle does not hold. This dissipation is exploited by engineers who design spillways to generate bores that remove destructive energy.

Bélanger originally used energy conservation, but obtained the wrong answer. He later corrected his analysis, applying the fundamental physical principles of mass and momentum conservation across the jump to obtain what we now call Bélanger's Equation, giving the ratio of water depths upstream and downstream.

You do not have to leave home to observe hydraulic jumps. They can be seen in the kitchen sink. When the water from the tap hits the sink pan, it flows radially and rapidly outwards in a thin layer until it slows to a critical speed where a circular hydraulic jump forms.

The sink bore is effectively a 'white hole'. In a black hole the gravitational attraction is so intense that light waves cannot escape, or pass out of the hole. A white hole has the opposite property: waves cannot enter the region within the jump.

Surface waves travel more slowly in shallow water than in deep water. In the shallow inner region of the sink bore the flow is more rapid than the wave speed. Surface waves are swept outwards by the flow and cannot propagate against it. Consequently, waves outside the hole cannot penetrate the inner region. They are trapped outside, just as light is trapped inside a black hole.

The water decelerates across the jump, where the flow speed and wave speed are equal and the depth increases by an order of magnitude. The jump is normally stationary but by turning the tap you can change the radius of the sink bore.

Hydraulic jumps continue to provide opportunities for engineers, sport for surfers and challenges for physicists and applied mathematicians. Who would have thought that bores could be so interesting?

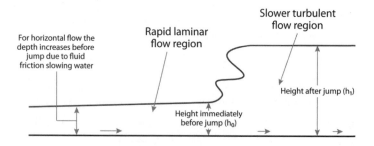

Characteristics of a hydraulic jump.

PYTHAGOREAN (OR BABYLONIAN) TRIPLES ‖

The Pythagorean theorem states that the square of the hypotenuse of a right-angled triangle is equal to the sum of the squares of the other two sides. It can be written as an equation,

$$a^2 + b^2 = c^2,$$

where c is the length of the hypotenuse, and a and b are the lengths of the other two sides. Jacob Bronowski called the theorem, which appears in Euclid, Book I Proposition 47, 'the most important theorem in the whole of mathematics'. It enables us to define distance s between two arbitrary points in the plane as

$$s^2 = (x_2 - x_1)^2 + (y_2 - y_1)^2$$

and leads on to the much more general metric of Riemannian geometry in curved space:

$$ds^2 = g_{mn} \, dx^m \, dx^n$$

Perhaps proved first by Pythagoras, it was known more than a thousand years earlier in Babylon, India and China.

The most familiar application of the result is the 3:4:5 rule, known for millennia to builders and carpenters. If a rope with knots spaced 1 metre apart is used to form a triangle with sides 3, 4 and 5 metres, the sides of length 3 and 4 meet at a right angle. This is the simplest example of a Pythagorean triple, $3^2 + 4^2 = 5^2$.

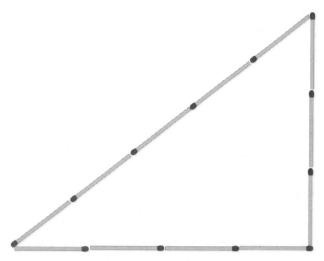

A right-angled triangle with sides of length 3, 4 and 5, made from 12 matches.

More generally, a Pythagorean triple is any set of any three *whole numbers* (a, b, c) that satisfy

$$a^2 + b^2 = c^2$$

giving three integral sides of a right triangle. If (a, b, c) is a Pythagorean triple, then so is (ka, kb, kc) for any integer k. If a, b and c have no common factors, the triple is called primitive. There are an infinite number of primitive triples.

A clay tablet originating from Mesopotamia around 1800 BC shows that the Babylonians were familiar with Pythagorean triples. Known as Plimpton 322, the tablet contains columns of numbers in cuneiform script (see picture below). These have been interpreted in terms of right-angled triangles: two of the columns list the largest and smallest elements of a Pythagorean triple or, in other words, a and c, the lengths of the hypotenuse and shortest leg of a right-angled triangle. Another column lists the square of the ratio of these values, effectively the squared cosecant of the angle opposite the shortest side. The order of the columns corresponds to increasing values of the angles, suggesting that the tablet may be an early trigonometric table, although this is controversial.

The tablet has 15 rows of numbers, but it is broken, and what additional entries might have been found on the original is unknown. John Conway and Richard Guy give a reconstruction with 34 rows, with sides corresponding to all regular numbers; that is, divisors of 60.

Diophantus of Alexandria showed that, for any whole numbers m and n, the three numbers

$$a = 2mn, b = n^2 - m^2, c = n^2 + m^2$$

form a Pythagorean triple. It is trivial to prove that $a^2 + b^2 = c^2$. Moreover, if mn is odd (m and n are of opposite parity) and m and n are coprime (no common factors) then the triple (a, b, c) is *primitive*, that is, a, b and c have no common factors.

We can tabulate the first few triples generated in this way:

m	n	$a = 2mn$	$b = n^2 - m^2$	$c = n^2 + m^2$
1	2	3	4	5
1	3	6	8	10
1	4	8	15	17
1	5	10	24	26
2	3	12	5	13
2	4	16	12	20
2	5	20	21	29
3	4	24	7	25
3	5	30	16	34
4	5	40	9	41
10	20	400	300	500

Nobody knows how the Babylonians calculated the values on the tablet, but they must have had a method something like the Diophantine algorithm.

SOURCES

Conway, John H. and Richard K. Guy (1996), *The Book of Numbers*. New York: Springer-Verlag.

Maor, Eli (2007), *The Pythagorean Theorem: A 4000-year History*. Princeton, NJ: Princeton University Press.

BÉZOUT'S THEOREM

Two lines in a plane intersect at one point, a line cuts a circle at two points, a cubic (an S-shaped curve) crosses the x-axis three times and two ellipses, one tall and one squat, intersect in four places.

In fact, *these four statements may or may not be true*. For example, two parallel lines do not meet, a line may 'miss' a circle, a cubic may have only one real root, or two ellipses may be distant and disjoint:

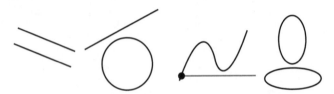

This is galling to the mathematician, who craves order and neatness. It is more satisfying to make statements that are unconditional or true in every case. We can rectify the situation if we introduce three new ideas.

1 COMPLEX NUMBERS

First we need to consider complex numbers, not just real numbers. The equation $y = x^2$ determines a parabola and $y = -1$ is a straight line. Eliminating y by substitution, we get $x^2 = -1$. Of course, this has no real solutions but it is satisfied by $x = \sqrt{-1} = i$ and $x = -\sqrt{-1} = -i$. We can say that the line intersects the parabola at the two imaginary points $(i, -1)$ and $(-i, -1)$. Similarly, a line that 'misses' a circle in the real plane actually intersects it at two imaginary points.

2 MULTIPLE ROOTS

But what happens if the line is tangent to the circle? We can look at this as the limiting case of two intersections as they move into coincidence. Thus, we regard a tangent contact as a double intersection and count it with multiplicity 2. And there are higher-order contacts, which must be counted with the appropriate multiplicity.

3 POINTS AT INFINITY

Finally, we need to add extra 'points at infinity' where two parallel lines meet. For each direction we have an additional point at infinity, but we regard the points at each extremity of a line as identical. This is quite difficult to visualise, but it leads to a consistent system called the projective plane. It is similar in a topological sense to a hemisphere in which opposite points on the rim are identified as equal.

BÉZOUT'S THEOREM

With these three embellishments, the four statements in the opening sentence are always true. Indeed, with these refinements, we have a much more general result called Bézout's Theorem. This states that any two curves defined by polynomial equations of degree m and n intersect in precisely mn points. We must realise that some of these intersections may coincide, some may have imaginary or complex coordinates and some may be on the 'line at infinity'.

The theorem of the French mathematician Étienne Bézout (1730–1783) has an appealing elegance and simplicity. It is a general statement, with no irksome exceptions or special cases. Under its conditions, algebraic curves are very well behaved. For two algebraic curves, the number of common points equals the product of the degrees of the two curves. For example, two cubic equations must intersect in nine points. What could be simpler?

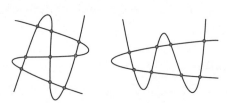

Left: Two cubics intersecting in 9 = 3 x 3 points. Right: a quadratic and quartic intersecting in 8 = 2 x 4 points.

The result of Bézout was known to Isaac Newton who, in his *Principia*, wrote that two curves have a number of intersections given by the product of their degrees. The proof was published by Bézout in 1779. The special case where one of the curves is a straight line reduces immediately to the Fundamental Theorem of Algebra: every non-zero, single-variable, degree n polynomial with complex coefficients has, counted with multiplicity, exactly n roots.

FRENCH CURVES AND BÉZIER SPLINES

A French curve is a template, normally plastic, used for manually drawing smooth curves. These simple drafting instruments provided innocent if puerile merriment to generations of engineering students, but they have now been rendered obsolete by computer-aided design (CAD) packages, which enable us to construct complicated curves and surfaces using mathematical functions called Bézier splines.

Bézier splines were first made popular by the automotive designer Pierre Bézier around 1962 when he was working for Renault. Bézier patented and popularised the use of these functions in computer-aided design, but he did not invent them.

An iterative binary algorithm for computing the curves had been devised some time earlier by another auto engineer, Paul de Casteljau, who was working at Citroën. The mathematical functions underlying them, introduced about half a century earlier by the Russian mathematician Sergei Bernstein, are known as Bernstein basis polynomials.

BERNSTEIN POLYNOMIALS

Sergei Bernstein (1880–1968) was born in Odessa. He was awarded a doctorate by the Sorbonne and worked for a year in Göttingen with David Hilbert. His mathematical work was on partial differential equations, differential geometry, probability theory and approximation theory. He introduced the Bernstein polynomials in 1912 when he gave a proof of the Weierstrass approximation theorem. This theorem states that any continuous function on a closed interval [a,b] can be uniformly approximated by polynomials, and Bernstein's proof was constructive: it gave explicit expressions for the approximating functions.

Expansions in Bernstein polynomials have a slow rate of convergence. For that reason, following Bernstein's application to the Weierstrass theorem, these polynomials had little impact in numerical analysis. But interest in them was reawakened in the 1960s, when Paul de Casteljau and Pierre Bézier showed their value in CAD.

BÉZIER SPLINES

Pierre Bézier sought a quantitative description of the geometry of car bodies, and mathematical tools that would allow designers to construct and manipulate complex shapes. He defined curves with control points that could be used like handles to intuitively shape a curve or surface to the desired form. This led him to the functions we now call Bézier curves. They were soon found to be expressed in terms of Bernstein polynomials.

The simplest Bézier curve is just linear interpolation between two control points, P_0 and P_1. The quadratic Bézier curve is a segment of a parabola between P_0 and P_2 such that the tangents at these end-points intersect at P_1. Most commonly used is the cubic Bézier curve. For this, the positions of the end-points are specified by P_0 and P_3, and the directions of the curve at these points can be determined by the control points P_1 and P_2.

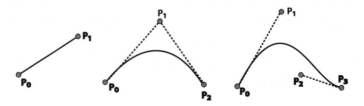

Cubic Bézier curves can be joined to form complex curves called Bézier splines, which pass smoothly through a series of points. In fact, the term 'spline' originates from a drafting instrument used by naval architects.

Bézier splines have a wide range of applications in computer graphics. They are powerful and intuitive to use in drafting diagrams and are used, for example, in the 'Draw' program of OpenOffice. They are also used to describe the character shapes of scalable fonts in text-processing programs like PostScript. Bézier splines are now indispensable tools for computer graphics, engineering design and animation.

SOURCES

Casselman, Bill (2005), *Mathematical Illustrations: A Manual of Geometry and PostScript*. Cambridge: Cambridge University Press.

Farouki, Rida T. (2012), 'The Bernstein polynomial basis: a centennial retrospective', *Computer Aided Geometric Design*, **29**, 379–419.

Joy, Kenneth I. (2000), 'Bernstein Polynomials: Online Geometric Modeling Notes', Department of Computer Science, University of California, Davis.

ASTRONOMICAL PERTURBATIONS

Remarkable progress in understanding the dynamics of the planets has been possible thanks to their relatively small masses and the overwhelming dominance of the sun. The sun is one million times more massive than earth. Jupiter, the largest planet, is three orders of magnitude smaller than the sun and three orders larger than earth. In a logarithmic sense, it is midway between earth and sun. The moon is two orders lighter than earth.

The vast differences in the masses of the planets allows us to simplify greatly the mathematical description of the solar system.

INTELLIGENT DESIGN OR NATURAL SELECTION?

The problem of two massive bodies, like the sun and earth, gravitationally attracting each other was solved completely by Newton. The orbits are elliptical, with the common centre of mass at a focus of the ellipse. With three or more bodies, the problem becomes vastly more complex, and is an ongoing field of research. It is only in very special cases that a solution is possible. But the configuration of the solar system is such that dramatic simplifications are possible.

It is almost as if a beneficent Creator had arranged for a configuration of the solar system that is within humankind's analytical abilities. It is by no means trivial to deduce the dynamics of the system, but the problem is not completely beyond our limited mental capacity. Of course, if the planets were larger, with more eccentric orbits, chaos would reign and the evolution of intelligent life would have been unlikely if not impossible.

TWO BODIES

Newton's law of gravitation gives the force between two bodies, of masses m_1 and m_2, separated by a distance r,

$$F = G\, m_1\, m_2\, /\, r^2$$

where G is the universal gravitational constant. This is the inverse square law of attraction. It is interesting that the force exerted by the earth on the sun is precisely equal in magnitude to that of the sun on the earth. However, recalling Newton's law of motion,

$$F = m\, a$$

we see that the *acceleration* of the sun is *one-millionth* that of the earth, since $m_{EARTH}/m_{SUN} \approx 10^{-6}$. Thus, we can treat the sun as stationary. Earlier, Kepler deduced from observations that the planetary orbits are ellipses with the sun at a focus. This is an excellent approximation to reality.

PERTURBATIONS

To some degree, each planet is attracted by all the others in addition to the sun's attraction. The main additional influence on the earth's trajectory is that of Jupiter. However, since Jupiter is never closer to earth than is the sun, and is one thousand times lighter than the sun, its gravitational attraction is minute compared to that of the sun. Thus we can treat the effect of Jupiter as a *small perturbation* of earth's elliptical orbit. This hugely simplifies the mathematical analysis.

Many other simplifications are possible thanks to the small deviations from ideal cases. Almost all the mass of the solar system is contained in the sun: the centre of mass of the system is deep within the solar sphere. Most orbits are close to circular, so the eccentricity may be treated as a small parameter. Moreover, all the orbits lie close to a plane, the ecliptic plane.

THE MOTION OF THE MOON

Lunar theory has challenged astronomers for centuries. Newton remarked that the motion of the moon was the one problem that caused him severe headaches. The relative attraction of the sun and earth on the moon is

$$(m_{SUN} / m_{EARTH}) \times (r_{EARTH}/r_{SUN})^2$$

$$\approx 10^6 \times (400{,}000 / 150{,}000{,}000)^2 \approx 7$$

Thus, the sun attracts the moon more strongly than does the earth. However, the accelerations of the earth and moon due to the sun are close to equal because they are at comparable distances from the sun. The difference is a small perturbation, which greatly facilitates the mathematical analysis: we can regard the earth–moon system as a single body orbiting the sun. Higher-order refinements come later!

Finally, we mention the case of an artificial satellite travelling between the earth and moon. Its mass is so tiny compared to that of the earth and moon that its gravitational effect on them is negligible. Once the motions of the earth and moon are known, the trajectory of the satellite can be calculated by treating it as a passive test-mass moving within their gravitational potential.

THE PREDICTIVE POWER OF MATHS

The mathematical equations that express the laws of physics describe phenomena seen in the real world. But they also allow us to anticipate completely new phenomena. Early in his career the Irish mathematician William Rowan Hamilton used the equations of optics to predict an effect called *conical refraction*, where light rays emerging from a biaxial crystal form a cone. This had never been seen before but, within a year, it was observed by his colleague Humphrey Lloyd, thrusting Hamilton into scientific prominence.

The essence of a good scientific theory is its predictive power. We develop the theory using observations of the world around us. Then we use it to account for new observations and also to predict entirely new phenomena. Prediction is the acid test of theory. The planet Neptune was mathematically predicted before it was directly observed. Newton's law of gravitation describes the motions of the planets around the sun. But, about two centuries ago, the orbit of Uranus was found to deviate from its predicted course.

What was wrong? Were the observations inaccurate, or were Newton's equations faulty? Neither of these; mathematical analysis showed that the orbital perturbations could be explained by another planet orbiting outside Uranus. In 1845 astronomers Urbain Le Verrier and John Couch Adams independently calculated the position of such a planet. Within a week, Neptune was found less than one degree from Le Verrier's computed location by an astronomer at the Berlin Observatory. This was a dramatic confirmation of Newton's theory of gravitation.

Le Verrier did not stop there. In 1859, he report that the slow precession of Mercury's orbit was not in agreement with Newtonian mechanics. But this time no new planet was found. The solution was more sensational: Newton's law of gravitation was imprecise. A new theory, Einstein's general relativity, was needed. This completely changed our view of space and time. It implied an orbit for Mercury in precise agreement with astronomical measurements. The new theory also predicted that the deflection of starlight grazing the sun's disc through an angle was double that expected from Newtonian theory. The observation of this effect during a solar eclipse in 1919 was a triumph for relativity theory.

Another prediction Einstein made in 1916 was that celestial bodies orbiting one another emit gravitational waves, ripples in the fabric of space–time. Indirect evidence of these waves was found in binary star systems that are spiralling inwards.

Further evidence emerged when astronomers at the Harvard–Smithsonian Center for Astrophysics released the 'first direct image of gravitational waves' in the infant universe. This evidence supports the theory of the Big Bang and cosmic inflation. However, solid confirmation of Einstein's prediction came only in early 2016, when scientists in the LIGO Project published dramatic observational evidence of gravitational waves originating from the collision of two black holes.

Another prediction is that radiating energy will eventually cause the earth to drop into the sun. Relax: this should not happen for many trillions of years. Long before that – within just a few billion years – the sun will become a red giant and swallow us up.

HIGHWAY GEOMETRY ‖

Next time you travel on a motorway, take heed of the graceful curves and elegant dips and crests of the road. Every twist and turn, every rise and sag, has been mathematically modelled to ensure that you have a pleasant and uneventful journey. Aesthetics plays a role in highway design, but the primary focus is on your safety and comfort.

The road alignment, or horizontal plan of the route, comprises straight sections and circular arcs, linked by smooth transitions to avoid any sudden changes. On a straight stretch or tangent section, there are no lateral forces. On a circular section there is an outward force, the centrifugal force, which varies with the square of the speed and with the curvature. The curvature, or bendiness, is the inverse of the radius, and is larger for sharper bends.

We are all familiar from school geometry with lines and circles, but the transitions between them are more exotic curves called *clothoids*. If the road were to change abruptly from a straight section to an arc, a sudden steering manoeuvre would be needed and a jerky onset of the centrifugal force would mean a very uncomfortable ride. To avoid this, a clothoid spiral section is interposed, linking the tangent to the arc.

The clothoid has a beautiful mathematical property: its curvature increases linearly with distance, from zero at the tangent end to the curvature of the arc where they meet. This means that the driver can turn the steering wheel gradually as she enters the curve, and the centrifugal force builds up slowly, minimising any discomfort.

We feel the centrifugal force strongly on a funfair ride, but on motorways it must be kept small. If the design speed of the road is high, wider arcs are needed to ensure smaller curvatures. Moreover, banking or 'camber' is added, with the road sloping downwards towards the inside to give a component of gravity that offsets the outward thrust.

The vertical profile of the road is also crucial for safety. It comprises steady grades – uphill, down or level – linked smoothly by parabolic sections. Here, visibility is a key concern: a sharp rise or dip can reduce driver vision drastically, especially at night, severely limiting the maximum safe speed. Engineers design roads to reduce these effects.

The calculation of the alignment and profile of motorways is done using complex computer programs. They generate realistic visual graphics, and the design engineer can view the road from the driver's perspective long before construction begins. The computer calculates the clothoid splines, parabolic profiles and more complex curves to ensure that the design meets exacting smoothness and safety standards.

Clothoid curves were first used to study diffraction, the bending of light around opaque objects. They are expressed in terms of mathematical quantities called Fresnel integrals, named for the French engineer Augustin-Jean Fresnel who helped to develop the theory of wave optics. But, as with many other mathematical innovations, clothoids were originally studied by Leonhard Euler, the Swiss genius who has been called 'the master of us all'.

Before the age of the motorway, clothoids were used to design transitions or easements on railway lines. More recently they have been employed in the construction of heart-stopping vertical loops on rollercoasters. Here the goal is not to reduce the centrifugal effect but to raise it to the highest safe level.

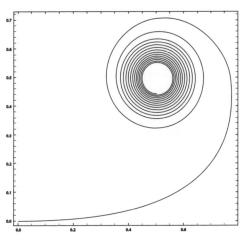

A cornu spiral has a clothoid form.

BREAKING WEATHER RECORDS

In arithmetic series, like $1 + 2 + 3 + 4 + 5 + \dots$, each term differs from the previous one by a fixed amount. There is a formula for calculating the sum of the first N terms. For geometric series, like $3 + 6 + 12 + 24 + \dots$, each term is a fixed multiple of the previous one. Again, there is a formula for the sum of the first N terms of such a series.

Another series arises when we take the inverse of each term in an arithmetic series. For the simplest arithmetic series, $1 + 2 + 3 + 4 + 5 + \dots$, the corresponding *harmonic series* is

$$1 + \tfrac{1}{2} + \tfrac{1}{3} + \tfrac{1}{4} + \tfrac{1}{5} + \dots \, .$$

This is more complicated. There is no general formula for the sum of N terms. We have to calculate it by hand or use a calculator or computer. As is well known, the series diverges, but very slowly.

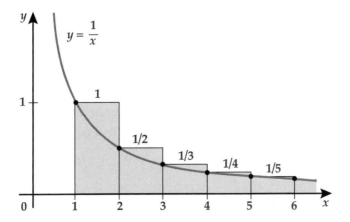

The shaded area between 1 and *n* gives the sum of the harmonic series to term (1/*n*).

BREAKING RECORDS

Harmonic series turn up in many contexts, in particular in the analysis of record-breaking values in a time series. Let us look at the mean annual temperature in Ireland. For now, let's forget about climate change and assume that the weather varies randomly but without any systematic drift or trend. How often, on average, would we expect to have a record high temperature?

If we have temperature data for only one year, it is automatically 'the hottest year on record' (as well as the coldest). If we now get data for the following year, there is a fifty–fifty chance that it is warmer than the first. The number of record years is now either 1 or 2. Taking the average, since we want a probability, we have 1 + ½. Now a third year's data comes in. There is a one in three chance that it is the hottest of the three, so the expected number of record-breaking years is now 1 + ½ + ⅓.

You can see where this is going: the expected number of record-breaking years in a series of mean temperatures for n years is 1 + ½ + ⅓ + ¼ + ⅕ + … + 1/n, the sum of the first n terms of the harmonic series. This number is called the nth harmonic number. Let us denote it by

$$H(n) = 1 + ½ + ⅓ + ¼ + ⅕ + ... + 1/n .$$

How does $H(n)$ behave as n gets bigger? If you calculate $H(n)$, you find that it grows ever more slowly with n. In fact, the sum diverges, but it creeps very gradually towards infinity. The first term is 1. The sum of the first ten is 2.9. With 100 terms we reach 5.2 and with 1000 only 7.5.

TEMPERATURE RECORDS

In ten years we would expect the temperature record to be broken about three times, since $H(10)$ = 2.9. In a century there would be about five new records, since $H(100)$ = 5.2. It is clear that records are more frequent in the early stages of the temperature record. However, in a steady-state climate, no matter how extreme the current record is, it is certain to be broken eventually, although the waiting time may be very long.

Of course, the real climate is changing and the mean temperature is rising inexorably. We need more than the simple harmonic series to deal with that!

THE FARADAY OF
STATISTICS ‖

In October 2012 a plaque was unveiled at St Patrick's National School, Blackrock, to commemorate William Sealy Gosset, who had lived nearby for 22 years during the early twentieth century. Sir Ronald Fisher, a giant among statisticians, called Gosset 'the Faraday of statistics', recognising his ability to grasp general principles and apply them to problems of practical significance. Gosset's name may not be familiar, but his work is known to anyone who has taken an introductory course in statistics. Using the pseudonym Student, he published a paper in 1908 that has been of importance ever since.

Gosset was born in Canterbury, Kent, in 1876 into an old Huguenot family. He studied chemistry and mathematics at Oxford, graduating in 1899 with first class honours in both subjects. He then joined Arthur Guinness & Son in Dublin as a chemist, and worked at the brewery in St James's Gate for 36 years, before becoming head brewer at a new Guinness brewery at Park Royal in London.

Guinness was interested in agricultural experimentation and hired scientists who could apply their expertise to the business. Gosset was one of these, and he used statistics to solve a range of problems connected with brewing, ranging from barley production to yeast fermentation, that affected the quality of the product. One problem involved the selection of varieties of barley that produced maximum yields in given soil types and allowing for the vagaries of climate.

To extend his knowledge, Gosset spent a year at the biometric laboratory of the leading statistician Karl Pearson at University College London. Reliable statistics require an adequate sample size. Gosset soon realised that Pearson's large-sample theory required refinement if it was to be useful for the small-sample problems arising in brewing. His fame today rests on a statistical test called Student's t-test.

But why Student? Gosset's main paper, 'The Probable Error of a Mean', was published in 1908. But to protect trade secrets, Guinness would not allow employees to publish the results of their research. They wished to keep secret from competitors the advantages gained from employing statisticians. Gosset persuaded his bosses that there was nothing in his work that would benefit competitors, and they allowed him to publish, but under an assumed name. Hence, anyone studying statistics encounters the name Student rather than that of the true author of the method.

Gosset's work has proved fundamental to statistical inference as practised today. His great discovery was to derive the correct distribution for the sample mean. Student's t-test arises when we estimate the average value of a randomly varying quantity from a small sample. It plays a crucial role in statistical analysis: for example, it is used to evaluate the effect of medical treatment, when we compare patients taking a new drug with a control group taking a placebo. It was also central to the development of quality control, which is vital in modern industry.

II THE CHAOS GAME

The term 'Chaos Game' was coined by Michael Barnsley, who developed this ingenious technique for generating mathematical objects called fractals. The Chaos Game is a simple algorithm that identifies one point in the plane at each stage. The sets of points that ultimately emerge from the procedure are remarkable for their intricate structure. *The relationship between the algorithm and fractal sets is not at all obvious* because there is no evident connection between them. This element of surprise is one of the delights of mathematics.

THE SIERPINSKI GASKET

Let's specify the algorithm in a simple case. We need a dice (I should say die but that feels odd) with three numbers. We can use a standard dice and identify 6 with 1, 5 with 2 and 4 with 3; that is, for face *n* we call it min $(n, 7 - n)$.

1. Fix three points in the plane, C_1, C_2 and C_3. For definiteness, we take the points $C_1 = (0, 0)$, $C_2 = (1, 0)$ and $C_3 = (0.5, \sqrt{3}/2)$, the corners of an equilateral triangle.

2. Pick any point P_0 and draw a dot. This is our starting point. At each stage, we denote the current point by P_k and call it the *game point*.

3. Roll the dice. If n comes up, draw a point halfway between P_k and C_n. For example, if we roll a 2, we pick the point halfway between the current point P_k and C_2. This is the new game point.

4. Repeat this procedure many times, drawing a new point at each step.

We seek the attractor of the mapping, so we eliminate the initial points, let's say the first 100 points. It is difficult to anticipate the outcome of this algorithm, but the figures below reveal all. We plot the results at several stages of the procedure, with $k = 500$, 1000 and 2000. We can see a structure forming. It is the famed Sierpinski Gasket.

The Chaos Game for three points at the vertices of an equilateral triangle, showing the output after 500, 1000 and 2000 steps.

BARNSLEY FERNS

Michael Barnsley generalised the Chaos Game to produce a wide range of fractal sets. He based it on a mathematical structure called an Iterated Function System (IFS). Starting with a point (x_k, y_k) in the plane, the next point is $f_r(x_k, y_k)$, where f_r is a member of the IFS, chosen with a specified probability p_r. In the simplest cases, the maps f_r are linear or affine transformations.

Probably the most famous example is the so-called Barnsley Fern. The algorithm or procedure is very similar to that described for the Sierpinski Gasket, but now four different linear maps are used instead of one, and the choices between them are made on the basis of random numbers, with prescribed probability for each map. The algorithm can be coded in a few lines of computer code. We show the results at several stages below.

Barnsley Fern generated by the Chaos Game. From left to right: 1000, 10,000 and 100,000 points.

The final result is a set of considerably beauty. The structure of the set built up by the Chaos Game bears a remarkable similarity to a real fern:

Barnsley Fern after one million iterations.

SOURCES

Barnsley, Michael (1993, 2000), *Fractals Everywhere*. San Diego, California: Academic Press.

Hutchinson, John (1980), 'Fractals and self-similarity', https://maths-people.anu.edu.au/~john/Assets/Research%20Papers/fractals_self-similarity.pdf. First published in *Indiana University Mathematics Journal*, **30**, 713–47.

Wikipedia, 'Chaos game', https://en.wikipedia.org/wiki/Chaos_game

Wikipedia, 'Barnsley fern', https://en.wikipedia.org/wiki/Barnsley_fern

FIBONACCI NUMBERS ARE GOOD FOR BUSINESS

Outside the entrance to the head office of the Irish Business and Employers Confederation (Ibec) in Baggot Street, Dublin, there is a plaque with a logo comprising a circular pattern of dots. According to Ibec's website, the logo brings 'dynamism' and hints at Ibec's 'member-centric ethos'. In fact, it is more interesting than this.

The logo is based on the spiral patterns found in many flowers and plants. Examining it we find 34 clockwise and 21 counter-clockwise spirals. These numbers are sequential entries in a famous number sequence called the Fibonacci sequence. This is no coincidence.

In 1202, mathematician Leonardo of Pisa, usually known as Fibonacci, published a book, *Liber Abbaci*, in which he described the sequence now known as the Fibonacci numbers. They are easily defined by an iterative process: starting with 0 and 1, each entry is the sum of the previous two. Thus, the sequence begins 0, 1, 1, 2, 3, 5, 8, 13, 21, 34, 55, 89 and 144.

As the values increase, the ratios of successive Fibonacci numbers – like 8/5, 13/8 and 21/13 – tend to a definite limit, about 1.618, called the *golden number*. It was known to the Greeks through their study of proportions and the geometry of the pentagon. If we divide a circle into two arcs whose ratio is the golden number, the shorter arc subtends an angle of about 137.5°. This is called the *golden angle*.

Distinctive spiral patterns are found in many plants. For example, the hexagons on pineapples fit together in interlocking families of helical spirals. The numbers of spirals are successive Fibonacci numbers. Sunflowers, which belong to the daisy family, usually have 55, 89 or 144 petals, and spiral patterns are evident in their seeds.

Biologists described long ago how phylla (leaves, petals, seeds, etc.) are arranged on plants: this branch of botany is called phyllotaxis. But explaining why these patterns form is much more difficult than describing them, and it is only recently that real progress has been made.

The seeds of a sunflower are arranged in a manner that makes efficient use of the available space, giving maximum room for each seed to flourish and minimising wastage of space. As a new seed sprouts forth at the growth tip of a plant, it naturally tends to grow where there is most open space. Each seed is displaced from the previous one by the golden angle.

But why the golden angle? Recent research shows that the angle emerges naturally as a feature of the dynamics of plant growth. Some years ago, Dublin-born Alan Newell, at the University of Arizona, applied elasticity theory to continuum models of growing cacti shoots. But mechanics could not answer all the questions.

Recently, Newell has shown that biochemistry, mechanics and geometry all play a role in generating the observed patterns. The growing seeds exert forces on each other, triggering the production of auxin, a growth-enhancing plant hormone. The solutions generated by the model of auxin concentration are found to be very similar to the patterns found in real flowers. This illustrates in a striking manner how nature is capable of producing optimal packing strategies.

BISCUITS, BOOKS, COINS AND CARDS: SEVERE HANGOVERS

Have you ever tried to build a high stack of coins? In theory it's fine: as long as the centre of mass of the coins above each level remains over the next coin, the stack should stand. But as the height grows, it becomes increasingly trickier to avoid collapse. In theory it is possible to achieve an arbitrarily large hangover. In practice, at more than about one coin diameter it starts to become difficult to maintain balance.

Let the coin radius be our unit of length. Assume the tilt is to the right and take the origin at the right-hand edge of the top coin. We build the stack downwards. To achieve the theoretical maximum overhang at each stage, we place the stack with its centre of mass just over the right-hand edge of the next coin. For n coins, let $C(n)$ be the horizontal distance from the origin to the centre of mass. Thus, the *overhang* $D(n + 1)$ for $n + 1$ coins is $C(n)$:

[Overhang for $n+1$ coins] = [Centre of mass for n coins]

If there is just one coin, $C(1) = 1$. With two coins, it is clear that the upper one may extend by up to one unit before it topples. The centre of mass is halfway between the centres of the two coins, 1.5 units from the origin. Thus, if the stack of two coins is placed on a third, the overhang is $D(3) = C(2) = 1.5$ units.

We proceed by induction to get the general result. The strategy is to place the balanced stack of coins on a new coin so that the augmented stack remains balanced. The centre of mass of the stack of n coins is a distance $C(n)$ from the origin. The centre of the new coin is displaced $C(n) + 1$ from the origin. The centre of mass of the $n + 1$ coins is the weighted average of the stack of n and the new coin:

$$C(n+1) = [n\,C(n) + (C(n) + 1)]/(n + 1) = C(n) + 1/(n + 1)$$

This is the maximum overhang for $(n + 1)$ coins. Thus we have the recursive relationship

$$C(n + 1) = C(n) + 1/(n + 1).$$

Since $C(1) = 1$, we have $C(2) = 1 + \frac{1}{2}$, $C(3) = 1 + \frac{1}{2} + \frac{1}{3}$ and in general

$$C(n) = H_n = 1 + \frac{1}{2} + \frac{1}{3} + \frac{1}{4} + \frac{1}{5} + \dots + 1/n$$

where H_n is the nth harmonic number.

The harmonic numbers increase without limit, but very slowly: the harmonic series is divergent. Thus, in principle, there is no limit to the extent of the overhang. However, the rate of increase of H_n with n is so slow that, in practice, only a limited overhang is possible.

The length unit for circular items like biscuits and coins is the radius. For rectangular items, like books and cards, the unit is half the length. The theoretical maximum overhang for ten volumes is $D(10) = C(9) = H_9$ units. This is about 1.4 book lengths.

With a deck of cards it is easier to achieve a large hangover. The image below shows a stack with an overhang of about 1.5 card lengths. The theoretical maximum for 52 cards is H_{51} units or $H_{51}/2$ card lengths, which is about 2.26 card lengths.

Try this yourself: you can probably do better!

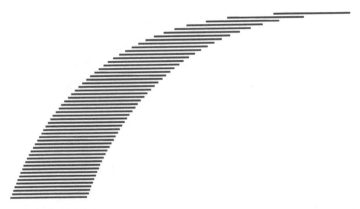

Schematic diagram of a stack of 52 cards with the theoretical maximum overhang.

GAUSS'S GREAT TRIANGLE AND THE SHAPE OF SPACE

In the 1820s Carl Friedrich Gauss carried out a surveying experiment to measure the sum of the three angles of a large triangle. Euclidean geometry tells us that this sum is always 180° or two right angles. But Gauss himself had discovered other geometries, which he called non-Euclidean. In these, the three angles of a triangle might add up to more – or less – than two right angles.

Gauss is considered by many to be the greatest mathematician who ever lived. Before the euro era, he was featured on the German 10 Deutschmark note. The front of the note shows Gauss and the normal probability curve, or bell curve, that he introduced and that is so fundamental.

The reverse side of the note shows a surveying sextant and a portion of the triangulation network in the vicinity of Hamburg that was constructed by Gauss in the course of his survey of the state of Hanover between 1821 and 1825.

THE GREAT TRIANGLE

Using his new invention, a surveying instrument called a *heliotrope*, Gauss took measurements from three mountains in Germany: Hohenhagen near Göttingen; Brocken in the Harz Mountains; and Inselberg in the Thüringer Wald to the south. In his survey of the kingdom of Hanover, Gauss had used these three peaks as 'trig points'. The three lines joining them form a great triangle with sides of length 69 km, 85 km and 107 km. The angle at Hohenhagen is close to a right angle, so the area of the triangle is about half the product of the two short sides, or about 3000 km^2.

Gauss assumed that light travels in a straight line. His sightings were along three lines in space. We should not confuse these with measurements along great circles on the curved surface of the earth, which would form a spherical triangle. Gauss was considering a plane triangle.

In Euclidean geometry, the sum of the three angles of every triangle is equal to 180°. In hyperbolic geometry there is an angular *deficit* so that the sum of the three angles is less than 180°. In spherical or elliptic geometry there is an *excess*: the angles add up to more than 180°. The magnitude of the excess or deficit grows with the area of the triangle.

CURVED SPACE

The geometry of physical space is a matter of measurement, and the character of space must be established by observations. For thousands of years it was assumed, on the basis of such observations, that Euclid's geometry is a faithful and precise representation of physical space. The word 'geometry' means measurement of the earth.

Euclidean space is flat: the quantity that measures the curvature of space is called the Riemann tensor. For Euclidean geometry, all components of this tensor vanish identically. It was taken for granted that physical space is flat, but general relativity has changed all that. Thanks to Einstein, we know that physical matter distorts the space around it; the Riemann tensor is non-zero and, near the planets and stars, space is curved.

The observed total angle found by Gauss was 180°, within the limitations of observational errors. If Gauss had been able to take measurements of sufficient accuracy, he might have found that the sum of the three angles of his great triangle was greater than two right angles by an amount

$$\varepsilon = (GM/Rc^2)*(A/R^2)$$

due to the mass M of earth (G is Newton's constant, c the speed of light, R the radius of the earth and A the area of the triangle. This correction comes to about 10^{-13} radians (there would be further, smaller, contributions from the sun and the other planets).

Was Gauss really observing the shape of space? There is no clear documentary evidence that Gauss was actually seeking evidence of non-Euclidean geometry of physical space. Indeed, doubt has been cast by some experts on this idea: mathematician John Conway pointed out that a departure from Euclidean geometry large enough to be measurable on the scale of the earth would result in enormous distortions on an astronomical scale, and would have been evident long before Gauss made his measurements. Moreover, Bühler, in his biography of Gauss, dismisses as a myth the idea that Gauss was considering the curvature of space. He considered that the purpose of the great triangle was to act as a control to check the consistency of the measurements of the smaller triangles within it.

Of course, the earth's intrinsic (spherical) curvature means that when we combine several small triangles to make a large one, there is a discrepancy. Gauss was well aware of this and indeed he calculated the magnitude of this small spherical effect.

We now know that the angular discrepancy attributable to relativistic effects is tiny, far below measurement errors. But Gauss would not have had knowledge of its magnitude. At the time of his survey, Gauss was immersed in the study of non-Euclidean geometry. He knew that the angular discrepancy grows with the area, so it is reasonable to suggest that he had spatial curvature in mind when he measured the great triangle, and that he might have thought it worthwhile to look for evidence of an angular discrepancy.

SOURCES

Bühler, W. K. (1981), *Gauss: A Biographical Study*. Berlin, Heidelberg, New York: Springer-Verlag.

Conway, John H. (1998), 'Gauss and the really large triangle', Math Forum, http://mathforum.org/kb/ message.jspa?messageID=1381049

Hartle, J. B. (2003), *Gravity: An Introduction to Einstein's General Relativity*. Boston, Massachusetts: Addison-Wesley.

DEGREES OF INFINITY ‖

At some point in their mathematical initiation, children often express bemusement at the idea that there is no biggest number. It is obvious that, given any number, one can be added to it to give a larger number. But the implication that there is no limit to this process is perplexing.

Indeed, the concept of infinity has exercised the greatest minds throughout the history of human thought. It can lead us into a quagmire of paradox from which escape seems hopeless. In the late nineteenth century, the German mathematician Georg Cantor showed that there are different degrees of infinity – indeed, an infinite number of them – and he brought to prominence several paradoxical results that had a profound impact on the subsequent development of the subject.

SET THEORY

Cantor was the inventor of set theory, which is a fundamental foundation of modern mathematics. A set is any collection of objects, physical or mathematical, actual or ideal. A particular number, say 5, is associated with all the sets having five elements. And, for any two of these sets, we can find a 1-to-1 correspondence, or bijection, between the elements of one set and those of the other. The number 5 is called the cardinality of these sets. Generalising this argument, Cantor treated any two sets as being of the same size, or cardinality, if there is a 1-to-1 correspondence between them.

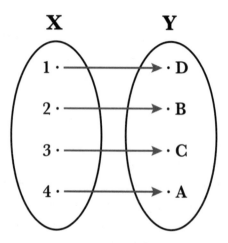

But suppose the sets are infinite. As a concrete example, take all the natural numbers, 1, 2, 3, ... as one set and all the even numbers 2, 4, 6, ... as the other. By associating any number n in the first set with $2n$ in the other, we have a perfect 1-to-1 correspondence. By Cantor's argument, the two sets are the same size. But this is paradoxical, for the set of natural numbers contains all the even numbers and also all the odd ones so, in an intuitive sense, it is larger. The same paradoxical result had been deduced by Galileo some 250 years earlier.

Cantor carried these ideas much further, showing in particular that the set of all the real numbers (or all the points on a line) have a degree of infinity, or cardinality, greater than the counting numbers. He did this by using an ingenious approach called the diagonal argument. This raised an issue, called the continuum hypothesis: is there a degree of infinity between these two?

INFINITIES WITHOUT LIMIT

Cantor introduced the concept of a power set: for any set A, the power set $P(A)$ is the collection of all the subsets of A. Cantor proved that the cardinality of $P(A)$ is greater than that of A. For finite sets, this is obvious; for infinite ones, it was startling. The result is now known as Cantor's Theorem, and he used his diagonal argument in proving it. He thus developed an entire hierarchy of transfinite cardinal numbers. The smallest of these is the cardinality of the natural numbers, called Aleph-zero:

Cantor's theory caused quite a stir; some of his mathematical contemporaries expressed dismay at its counter-intuitive consequences. Henri Poincaré, the leading luminary of the day, described the theory as a 'grave disease' of mathematics, while Leopold Kronecker denounced Cantor as a renegade and a 'corrupter of youth'. This hostility may have contributed to the depression that Cantor suffered through his latter years. But David Hilbert championed Cantor's ideas, famously predicting that 'no one will drive us from the paradise that Cantor has created for us.'

A SWINGING WAY TO SEE THE SPINNING GLOBE

Spectators gathered in Paris in March 1851 were astonished to witness visible evidence of the earth's rotation. With a simple apparatus comprising a heavy ball swinging on a wire, Léon Foucault showed how the earth rotates on its axis. His demonstration caused a sensation, and Foucault achieved instant and lasting fame.

The announcement of the experiment read, 'You are invited to see the Earth spinning ...' A bob of mass 28 kg was suspended from the dome of the Panthéon by a 67-metre wire. The pendulum swung through a diameter of about six metres, and the position was indicated on a large circular scale. Demonstrations were held daily, and attracted large crowds.

Following Foucault's demonstration, pendulum mania raged across Europe and the United States, and the experiment was repeated hundreds of times. Many of these attempts were done without due care. The *London Literary Gazette* reported on several cases in which 'to the horror of the spectators, the earth has been shown to turn the wrong way'. These errors probably resulted from elliptical bob trajectories due to incorrect starting conditions or to stray air currents disturbing the movement of the bob.

The observed change in the swing plane of the pendulum is often stated to be due to the earth turning beneath it. This is roughly correct, but it is an over-simplification. The turning rate for a pendulum swinging at the North Pole is one revolution per day. At other locations the rate depends on the latitude and at the equator there is no turning at all. Thus, after one day, the swing plane does not return to its original position. At Paris the turning period is about 31 hours. The mathematical term for this phenomenon is 'anholonomy', and it has been a source of confusion ever since Foucault's demonstration.

In September 1851, the *American Journal of Science* surveyed several pendulum demonstrations in Europe and America. It included details of the experiments carried out in Dublin by Galbraith and Haughton. The two reverends, Joseph Galbraith and Samuel Haughton, close contemporaries and lifelong collaborators, were both Fellows of Trinity College Dublin. They were well known for their many mathematical textbooks, which were widely used and which earned them handsome royalties.

Galbraith and Haughton replicated the pendulum experiment shortly after Foucault had reported his findings. Their experiment was done at the engine factory of the Dublin and Kingstown Railway, where Samuel's cousin Wilfred Haughton was chief engineer. They also analysed the effects of ellipticity of the trajectory and derived a mathematical expression for the precession due to this effect.

The pendulum length for the experiments of Galbraith and Haughton was 35 feet and its swing length 4 feet. The bob was an iron sphere, 30 lb in weight, with a downward-pointing spike to indicate its position on a scale. The theoretical precession rate at the latitude of Dublin is 12.07 degrees per hour. The mean rate observed in the experiments was 11.9 degrees per hour, which is surprisingly accurate considering the many possible sources of error. According to an article in the *Philosophical Magazine*, 'Messrs. Galbraith and Haughton ... have pursued their research with all imaginable precautions.' Their impressive results confirm this view.

DO YOU REMEMBER VENN?

Do you recall coming across those diagrams with overlapping circles that were popularised in the sixties, in conjunction with the 'New Maths'? They were originally introduced around 1880 by John Venn, and now bear his name.

Venn was a logician and philosopher, born in Hull, Yorkshire in 1834. He studied at Cambridge University, graduating in 1857 as sixth Wrangler in the Mathematical Tripos; that is, sixth best in mathematics in the entire university that year. He was ordained five years later, having come from a family with a long tradition of producing churchmen. His was the eighth generation of the family to have a university education. Venn introduced the diagrams in his book *Symbolic Logic*, published in 1881.

The idea of a set is among the most fundamental concepts in mathematics. A set is any well-defined collection of distinct objects. These objects are called the members or *elements* of the set. They may be finite or infinite in number. Set theory was founded by the

German mathematician Georg Cantor (see page 181), who discovered many remarkable and counter-intuitive properties of infinite sets.

Venn diagrams are very valuable for illustrating elementary properties of sets. They usually comprise a small number of overlapping circles; the interior of a circle represents a collection of numbers or objects or perhaps some more abstract set.

We often draw a rectangle to represent the 'universe', the set of all objects under current consideration. For example, suppose we consider all species of animals as the universe. The diagram below represents this universe, and the two circles indicate subsets containing animals of two separate species. The aggregate of all the elements of the two sets is called their *union*.

The elements that are in both sets make up the *intersection*. If one set contains all two-legged animals and the other has all flying animals, then bears, bees and birds are in the union, but only birds are in the intersection.

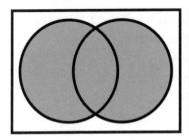 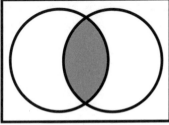

In the diagram on the next page, the elements of our universe are all the people from Connacht. We see three subsets shown by circles: redheads, singers and left-handers, all from Connacht. Clearly, these sets overlap and, indeed, there are some copper-topped, crooning cithogues in Connacht, located in the central shaded region where the three circles overlap.

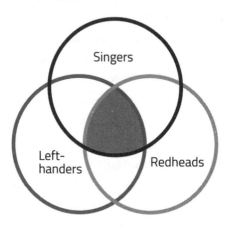

The three overlapping circles have attained an iconic status, and are used in a huge range of contexts. It is possible to devise Venn diagrams with four or more sets, but they are not as popular because the regions can no longer all be represented by circles if all possible combinations of membership are allowed for, and the essential simplicity of the three-set diagram is lost.

MATHEMATICS IS COMING TO LIFE IN A BIG WAY ‖

Once upon a time biology meant zoology and botany, the study of animals and plants. The invention of the microscope shifted the emphasis to the level of cells, and more recently the focus has been at the molecular level. Biology has been transformed from a descriptive science to a quantitative discipline; mathematics now plays a vital role, and students of biology need mathematical skills.

Biological systems are hugely complex, but simple mathematical models can isolate and elucidate key elements and processes and predict crucial aspects of behaviour. Many problems in biology have been solved using mathematics already developed in other fields – network analysis, group theory, differential equations, probability, chaos theory and combinatorics – but completely new mathematical techniques may be required to solve some tough problems in the life sciences.

The shape of a protein is an essential factor in determining its functions. For example, haemoglobin has a complex folded shape that enables it to 'pick up' an oxygen molecule and 'drop it' where it is needed. The folding and tangling of protein molecules is being modelled using the branch of topology called knot theory.

Network analysis shows us that a large network of simple components – whether transistors or neurons – can exhibit astonishingly complex behaviour. The human brain has 100 billion nerve cells, linked together by a biological wiring system of axons and dendrites. The number of interconnections is vast, something like a thousand million million. This is 'big data' with a vengeance.

One simple element can do little: link a large number together and you can get fantastically complex behaviour. For the brain, this includes thinking! Many questions in neuroscience remain to be answered: How does memory work? How is information from the senses interpreted and stored? Bio-informatics deals with the enormous data sets produced in biological research.

Systems biology is a rapidly developing inter-disciplinary field of biological research. It focuses on complex inter-actions using a holistic approach aimed at discovering and understanding emergent properties of organisms. Such properties are difficult or impossible to understand using a reductionist approach – breaking them down into basic constituents. Systems biology makes extensive use of mathematical and computational models.

The communication networks in the human body involve millions of interlinked cells. Occasionally, these networks break down, causing diseases like cancer. Systems Biology Ireland (SBI) at UCD is designing new therapeutic approaches based on a systems-level, mechanistic understanding of cellular networks. Researchers at SBI apply mathematics and computer science to enormous data sets arising from biological techniques. Their research aims to find out what genes do, how they work together, what goes wrong in diseases, and how to cure them.

Just as astronomy gave rise to spectacular developments in mathematical analysis in the eighteenth century, biology may have a profound effect on mathematics in the future. Some commentators see bio-mathematics as the great frontier of the twenty-first century, and argue that by 2100 biology and mathematics will have changed each other dramatically, just as mathematics and physics did in earlier centuries.

TEMPERAMENTAL TUNING

Every pure musical pitch has a frequency – the number of oscillations per second in the sound wave. Doubling the frequency corresponds to moving up one octave. A musical note consists of a base frequency or pitch, called the *fundamental*, together with a series of harmonics, or oscillations, whose frequencies are whole-number multiples of the fundamental frequency.

Each musical instrument has a different pattern of harmonics that give it an individual quality or character. Thus, the note A sounds quite distinct when played on an oboe and a clarinet: the fundamentals are the same, but the overtones or harmonics are different.

In Western music, the octave, ranging from a *tonic* note to one of twice the frequency, has twelve distinct notes, the interval between successive notes being a semitone. There are many ways to determine the precise frequencies of the notes, no single scheme being ideal – the best scheme depends on the type of music.

PYTHAGOREAN TUNING

Pythagoras discovered that a perfect fifth, with a frequency ratio of 3:2, is especially consonant. The entire musical scale can be constructed using only the ratios 2:1 (octaves) and 3:2 (fifths). Our perception of musical intervals is logarithmic, so adding intervals corresponds to multiplying frequency ratios.

In the tonic sol-fa scale the eight notes of the major scale are Do, Re, Mi, Fa, So, La, Ti, Do. Starting with 'Do', the ratio 3:2 brings us to 'So'. Moving up another fifth, we have the ratio 9:4. Reducing this by 2 to remain within the octave, we get 9:8, the note 'Re'. Moving up another fifth gives a ratio 27:16 and brings us to 'La'. Continuing thus, we get all the notes in the major scale. This is called Pythagorean tuning:

Do	Re	Mi	Fa	So	La	Ti	Do
1:1	9:8	81:64	4:3	3:2	27:16	243:128	2:1

The Pythagoreans noticed that $2^{19} \approx 3^{12}$, so going up twelve fifths, with ratio $(3/2)^{12}$ and down seven octaves $(1/2)^7$ gets you back (almost) to your starting point. The number $3^{12}/2^{19} \approx 1.01364$ is called the Pythagorean comma. The frequencies of *enharmonics*, such as F sharp and G flat, differ by this ratio.

CIRCLE OF FIFTHS

The Circle of Fifths is a diagram representing the relationship between musical pitches and key signatures.

It is a geometric diagram showing the twelve notes of the chromatic scale. The circle is useful in harmonising melodies and in building chords.

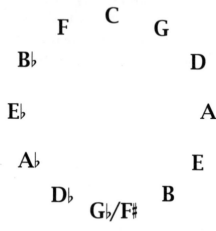

We start at the top with middle C, the base note of the scale of C major, which has no sharps or flats. Moving clockwise to one o'clock we have G (1 sharp), then D (2 sharps) and so on to F sharp (which has six sharps) at 6 o'clock. Proceeding counter-clockwise from the top, we have F (1 flat), B flat (2 flats) and onward to G flat (six flats) at 6 o'clock. The notes F sharp and G flat are called *enharmonics*.

TRIADS AND JUST INTONATION

The triad – three notes separated by four and three semitones – such as C–E–G, is of central importance in western music. In the tuning scheme of Pythagoras, the third (C–E) has a frequency ratio of 81:64. Generally, ratios with smaller numbers result in more pleasant sensations of sound. A more consonant sound is given by replacing 81:64 by 80:64 = 5:4. The three notes of the triad C–E–G are then in the ratio 4:5:6. Likewise, changing the sixth note (A or 'La') from 27:16 to 25:15 = 5:3 makes F–A–C a perfect triad with the frequency ratios 4:5:6. Finally, if the monstrous 243:128 is replaced by 240:128 = 15:8, we get a scheme of tuning called *just intonation*:

Do	Re	Mi	Fa	So	La	Ti	Do
1:1	9:8	5:4	4:3	3:2	5:3	15:8	2:1

TEMPERED SCALES

It is impossible to tune an instrument like a piano so that all fifths have perfect frequency ratios of 3:2. There are tuning schemes that make small compromises so that all ratios are close to the perfect values, and modulation or change between one key and another is possible. In such *tempered* scales, the ratios are all slightly imperfect, but close enough to be acceptable to the ear.

In Western music it is essential to be able to change smoothly between keys without dissonant consequences. The scheme called equal temperament achieves this. The idea is to make all semitone intervals equal. There are twelve semitones in an octave, and adding intervals corresponds to multiplying frequency ratios. Since an octave has ratio 2:1, we need a number that yields 2 when multiplied by itself 12 times. This number is just the twelfth root of 2 or $^{12}\sqrt{2} \approx 1.059$ and it is the key (!) to modern tuning. In this system, no harmonic relationships are perfect, but all are acceptable to the ear.

In equal temperament, the twelve steps all have identical frequency ratios. There are no enharmonics: F sharp and G flat are identical. In the perfect harmonic system, they are distinct notes. Equal temperament is a compromise, and there are those who argue that it is *like black and white compared to the colour of just intonation*. The best way to decide is to listen.

SOURCES

Benson, D. J. (2007), *Music: A Mathematical Offering*. Cambridge University Press.

Harkleroad, Leon (2006), *The Math Behind the Music*. Cambridge University Press.

CARTOON CURVES ‖

The powerful and versatile computational software program called Mathematica is widely used in science, engineering and mathematics. There is a related system called Wolfram Alpha, a computational knowledge engine, that can do Mathematica calculations and that runs on an iPad.

The bear curve. The Mathematica command to generate this is given on page 201.

Mathematica can do numerical and symbolic calculations. Algebraic manipulations, differential equations and integrals are simple, and a huge range of beautiful graphs can be plotted in an instant.

Searching in Wolfram Alpha, you can find some graphs called 'heart curves', one of which has the equation

$$(x^2 + y^2 - 1)^3 - x^2 y^3 = 0$$

Heart-shaped curves arise in some applications such as the Werner Projection in cartography. The algebraic form of the heart curve is a sixth-order polynomial in x and y. Note that since only even powers of x occur, the curve has bilateral symmetry, a property found in the bodies of many animals and, of course, in the human face.

We can change the right-hand side of the equation from zero to a value α:

$$(x^2 + y^2 - 1)^3 - x^2 y^3 = \alpha$$

The graph changes gradually from the heart curve to a circle as α increases to larger positive values. However, as α becomes more negative the curve changes in more interesting ways. It develops lobes at the top left and right and eventually, at about $\alpha = -0.1$, these break off into separate closed curves. For $\alpha < -1.0$ there are no real points on the curve and the plot is empty.

Plots of the heart curve $(x^2 + y^2 - 1)^3 - x^2 y^3 = \alpha$ for $\alpha = \{0.1, 0.0, -0.1, -0.2, -0.5\}$.

A nice 'fun curve' results for $\alpha = -0.1$ when the lobes have developed to look like the ears of a cartoon character (see middle plot opposite). In the plot at the top of this article, we combine two plots, one with $\alpha = -0.1$ for the outline and one with $\alpha = -0.2$, but scaled down to half size, for the eyes and nose. In fact, this 'bear curve' can be generated with a single command in Mathematica:

ContourPlot [((x^2 + y^2 - 1)^3 - x^2 y^3 + 0.1) * (((2 x)^2 + (2 y)^2 - 1)^3 - (2 x)^2 *(2 y)^3 + 0.2),{x, -1.5, +1.5}, {y, -1.25, +1.75}, Contours -> {0},

ContourShading -> True, ContourStyle -> Thickness [0.01], FrameTicks -> False, Axes -> None, Frame -> False, ImageSize -> Large].

HOW BIG WAS THE BOMB?

Physicists, engineers and applied mathematicians have an arsenal of problem-solving techniques. Computers crunch out numerical solutions in short order, but it is vital to be able to verify these answers. The first 'reality check' is to confirm that the solution is roughly the right size. The physicist Enrico Fermi was brilliant at 'back of the envelope' calculations. He could estimate the size of the answer quickly and painlessly with a few simple calculations.

Another essential check is to ensure that formulas have the correct physical dimensions. The three fundamental dimensions in mechanics are length, mass and time. Other quantities are combinations of these. Area is length squared and volume is length cubed. Density is mass per unit volume, velocity is length per unit time, and acceleration, which is the rate of change of velocity, is length over time squared.

Every physical formula must be dimensionally consistent. If a length is expressed as a combination of quantities, the dimension of that combination must be a length. But dimensional analysis is much more fruitful that this.

Let us take a simple example. A brick is dropped from a tall building. How long does it take to reach the ground? The relevant parameters are the mass of the brick, the height of the building and the acceleration of gravity. To produce a quantity with dimensions of time, we divide the height by the acceleration of gravity and take the square root. This gives an estimate of the fall-time of the brick, correct except for a factor of root 2. The time is independent of the mass, as spotted by Galileo.

Geoffrey Ingram Taylor was one of the great scientists of the twentieth century. His grandfather was George Boole, first professor of mathematics at Queen's College Cork, now University College Cork. Taylor was a meteorologist in his early career and became a major figure in fluid dynamics, his ideas having wide application in meteorology, oceanography, engineering and hydraulics.

The first atomic blast, the Trinity Test in New Mexico, had an explosive yield of about 20 kilotons, the energy released by detonating 20,000 tons of TNT. But this value was secret: following World War II, details of atomic weapons remained classified. In 1947, photographs of the Trinity Test were released by the US Army. They aroused great interest and appeared in newspapers and magazines all over the world.

Taylor realised that these photographs would allow him to estimate the energy of the explosion. The images were taken at precise time intervals following the initial blast, and the horizontal scale was also indicated. Taylor used the technique of dimensional analysis. He assumed that the radius of the expanding fireball should depend only on the energy, the time since detonation and the density of the air. He combined these quantities to get an expression with the dimensions of energy, and found that it was proportional to the fifth power of the blast radius.

Taylor's value for the energy was 17 kilotons, close to the value later announced by President Truman. Taylor's result caused quite a stir and he was 'mildly admonished' by the US Army for publishing his deductions. But it was a mathematical tour de force. Considering the many sources of error, the estimate was, in his own words, 'better than the nature of the measurements permits one to expect'.

ALGEBRA IN THE GOLDEN AGE ‖

Some years ago a book was published with the title *How the Irish Saved Civilisation*. It told of how, by copying ancient manuscripts, Irish monks preserved the knowledge of ancient times through the Dark Ages, when Europe was plunged into ignorance and turmoil. Far to the East, a more extensive rescue operation was under way.

Evidence of Arabic science, the science practised by the scholars of the Islamic Empire, is all around us. For example, most of the stars in the sky have Arabic names. We owe a huge debt to the achievements of the scholars of the Arabic world. The scientific revolution in sixteenth- and seventeenth-century Europe depended greatly on the advances made in mathematics, physics, chemistry, medicine and philosophy in the medieval Islamic world.

Following the arrival of Islam in the early seventh century, an empire with a flourishing civilisation emerged, the Golden Age of Islam. During this period, Arabic was the language of international science and mathematics over a region stretching from India to Spain.

After the foundation of Baghdad in AD 762, an extensive programme of translation was started, in which the great works of India, Persia and Greece were translated into Arabic. Baghdad became a centre of enlightenment, culture and learning, with the House of Wisdom, established by Caliph al-Ma'mun, playing a role similar to the earlier Library of Alexandria.

One of the greatest scholars of the Golden Age was the Persian mathematician Muhammad ibn Musa al-Khwarizmi. He worked in the House of Wisdom around AD 820, where he wrote several astronomical and mathematical texts. One of his works, giving an account of the decimal system of numbers, is based on the earlier work of the Indian mathematician Brahmagupta. After al-Khwarizmi's book was translated into Latin, the decimal system of numbers was popularised in Europe by Leonardo of Pisa, also known as Fibonacci, in his *Liber Abbaci*.

Al-Khwarizmi's greatest work was *Kitab Al-Jebr*, translated as *The Compendious Book on Calculation by Completion and Balancing*. In this book, quadratic equations are solved by the process of completing squares, the method taught in schools to this day. The term 'al-jebr' in the title gives us the word algebra. The text played for algebra a role analogous to that played by Euclid's *Elements* for geometry: it was the best text available on the subject until modern times and it gave a new direction to the development of mathematics.

There were good reasons for developing new mathematical methods, which, according to al-Khwarizmi, 'men constantly require in cases of inheritance, legacies, partition, lawsuits and trade'. He also considered practical problems such as measuring land and digging canals.

Had it not been for the translation programme during the Golden Age of Islam, we would not have access today to many mathematical texts of the Greeks. Moreover, the knowledge transferred to Europe in the Renaissance was richer in many ways than that inherited by the Arabs from Greece, Persia and India.

The unique collection of the Chester Beatty Library includes some 2650 Arabic manuscripts. They cover religion, geography, medicine, astrology, mathematics and more, with original works and translations from ancient Greek. However, a complete history of Arabic mathematics is not yet possible since many manuscripts remain unstudied.

OLD OCTONIONS MAY RULE THE WORLD

On 16 October 1843, the great Irish mathematician William Rowan Hamilton discovered a new kind of number called *quaternions*. Each quaternion has four parts, like the coordinates of a point in four-dimensional space. Physical space has three dimensions, with coordinates (x, y, z) giving the east–west, north–south and up–down positions.

Quaternions can be added, subtracted, multiplied and divided like ordinary numbers, but there is a catch. Two quaternions multiplied together give different answers depending on the order: A times B is not equal to B times A. They break the algebraic rule called the commutative law of multiplication.

When Hamilton told his friend John T. Graves about the new numbers, Graves wondered 'Why stop at four dimensions?' He soon came up with a system of numbers, now called *octonions*, each of which has eight components: they are points in eight-dimensional space. But there was a price to pay: the octonians break another algebraic rule, the associative law.

Hamilton promised Graves that he would speak about the octonions at the Royal Irish Academy, but he was so excited by his quaternions that he forgot, and the English algebraist Arthur Cayley got the credit when he rediscovered the eight-component numbers a few years later. Only then did Hamilton bring news of Graves' work to the Academy.

You might wonder – just as Graves did – whether more new numbers can be found. A number system in which we can add, subtract, multiply and divide is called a *division algebra*. It turns out that there are division algebras only for dimensions 1, 2, 4 and 8. This was long believed to be true but was not proved until 1958.

Quaternions fell out of favour after Hamilton's death and were supplanted by vector calculus, which was more efficient and less complicated. They have resurfaced recently in computer graphics and astronautics, and have applications in theoretical mechanics.

Octonions fared even worse than quaternions, being regarded as little more than interesting curiosities. But now things may be changing: for decades, physicists have been struggling to reconcile quantum mechanics and general relativity. The goal is to produce a unified description of nature, sometimes called a Theory of Everything. The foremost candidate is called string theory: elementary constituents like electrons are not point particles, but one-dimensional oscillating strings that trace out sheets or tubes as they move through time.

String theory depends on the idea of *supersymmetry*: every matter particle has a 'twin' force particle, and the laws of physics remain unchanged if we swap the matter and force particles. Currently, quantum mechanics represents the two species by different kinds of numbers, spinors and vectors. If we could represent both species by the same kind of numbers, we would have a symmetric description of matter and force, that is, supersymmetry. This is possible only in one, two, four or eight dimensions.

Adding two more components, one for time and one for the string dimension, we get spaces of three, four, six and ten dimensions. At present, ten-dimensional space based on octonions is considered by string theorists to be a likely candidate for describing the universe. This is speculative and controversial, but if it turns out to be correct, matter and force particles will be modelled by those curious eight-component octonions discovered by Graves more than 170 years ago.

x	i	j	k	E	I	J	K
i	-1	k	$-j$	I	$-E$	$-K$	J
j	$-k$	-1	i	J	K	$-E$	$-I$
k	j	$-i$	-1	K	$-J$	I	$-E$
E	$-I$	$-J$	$-K$	-1	i	j	k
I	E	$-K$	J	$-i$	-1	$-k$	j
J	K	E	$-I$	$-j$	k	-1	$-i$
K	$-J$	I	E	$-k$	$-j$	i	-1

Multiplication table for octonions, of the form
$z = a + bi + cj + dk + eE + fI + gJ + hK$.

LIGHT WEIGHT II

A star near the limb of the sun appears to be slightly displaced outwards radially compared to its position when the sun is absent. A remarkable series of observations during the solar eclipse in May 1919 confirmed that light passing close to the sun is deflected by its gravitation. Albert Einstein had calculated, based on his general theory of relativity, that the deflection of a sun-grazing beam would be 1.75 seconds of arc. This is a small angle but it was well within the then-current techniques of astronomy.

Two expeditions set out from Britain, one to the island of Principe, lying off the west coast of Africa, the other to Sobral in northern Brazil, both on the path of totality of the eclipse. The observations at Principe gave an estimated deflection of 1.61+/−0.30 arc seconds, in excellent agreement with Einstein's prediction. The Sobral results provided further confirmation. This experiment catapulted Einstein into world fame and he has remained an icon of science ever since.

It is often stated that the bending of light is a purely relativistic phenomenon, but this is not correct. Classical Newtonian mechanics predicts such a deflection, but with half the magnitude of that observed in 1919. Newton himself considered that light has weight and should be influenced by the force of gravity. In his *Opticks*, Newton writes: 'Do not Bodies act upon Light at distance and by their action bend its Rays?'

Newton supported the corpuscular theory of light, regarding it as comprising particles with small but finite mass. He concluded that such particles would be influenced by a gravitational field.

It is a surprising fact that the motion of a test particle orbiting a massive body is independent of the mass of the particle. By Newton's laws, the motion of the test mass is governed by the equation

$$m\,a = G\,M\,m\,/\,r^{\,2}$$

where m is the test mass, a is its acceleration, M the solar mass, r the distance and G is Newton's gravitational constant. On the left, m represents the inertial mass; on the right it represents the gravitational mass of the particle. But, owing to the equivalence of inertial and gravitational mass, they cancel out, so the motion does not depend on the mass of the particle.

We now consider an asteroid approaching the sun from a great distance. For small values of the energy, the asteroid follows an elliptical orbit, returning repeatedly to its initial position. We assume that the asteroid has high energy, so that it traces out a hyperbolic orbit, with the equation

$$x^2 / a^2 - y^2 / b^2 = 1$$

where a and b are called the major and minor semi-axes of the hyperbola.

Let V be the limiting speed of the asteroid when it is remote from the sun and moving along a line whose perpendicular distance from the sun is R (see the diagram).

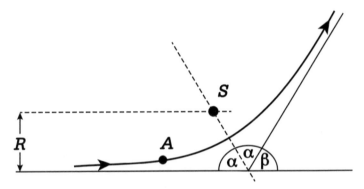

Trajectory of an asteroid A moving past the sun S on a hyperbolic orbit.

Newton obtained the bending angle for a ray of light:

$$\beta = 2GM/Rc^2$$

Using the known values for the constants, Newton's theory predicts that starlight passing close to the sun will be bent by an angle of 0.87 arc seconds.

Clearly, the astronomical observations in Principe and Sobral could not be reconciled with Newton's theory, and a scientific revolution ensued. Relativity showed that there is an additional effect, the distortion of spacetime due to the presence of the sun. The details of this are found in Einstein's original paper on general relativity (1916). The final result is that the bending angle is

$$\beta = 4GM/Rc^2 \approx 0.847 \times 10^{-5} \text{ radians} = 1.75 \text{ arc seconds}$$

which is twice the value of the deflection angle predicted by classical mechanics. The astronomical observations in 1919 were a splendid confirmation of this prediction.

FALLING BODIES   II

Aristotle was clear: heavy bodies fall faster than light ones. He arrived at this conclusion by pure reasoning, without experiment. Today we insist on a physical demonstration before such a conclusion is accepted. Galileo tested Aristotle's theory: he dropped bodies of different weights simultaneously from the Leaning Tower of Pisa and found that, to a good approximation, they hit the ground at the same time.

Galileo contrived a thought experiment, or *Gedankenexperiment*, to show that heavy and light weights must fall at the same rate. For suppose the case is otherwise. Then, if the two weights are linked by a string, the heavy one will tug the light one after it and the light one will retard the heavy one. But the two coupled weights can be regarded as a single one which, being heavier than either, should fall even faster. This contradiction leads us to conclude that the initial supposition is false.

But Galileo was not content with such reasoning: he sought to demonstrate his conclusion physically. His pupil Vincenzo Viviani wrote that in 1589 Galileo dropped balls of the same material but of different sizes from Pisa's tower to show, 'to the dismay of the philosophers', that they fell at the same speed, contrary to Aristotle. As no written record by the master himself is extant, spoil-sport historians cast doubt on this story. However, a recent authoritative biography points out that the tilt of the tower made it a perfect platform for the experiment.

AIR RESISTANCE AND SKY-DIVING

In fact, the drag of the air causes a small difference in fall rate, so wooden and iron balls of the same size will hit the ground at slightly different times. Galileo was aware of this effect, which is minute compared to Aristotle's idea of fall rate proportional to weight. We all know that a hammer and a feather dropped together will not fall at the same rate. But if there is no air resistance, they will hit the ground together, as was demonstrated on the moon by astronaut David Scott on the Apollo 15 mission.

A sky-diver plummets towards the earth, accelerated by gravity but impeded by air resistance. His terminal velocity V_T is that rate of fall is such that the upward drag force due to air resistance exactly balances the downward pull of gravity (for simplicity, we neglect buoyancy).

At the terminal velocity, there is no net force, so the sky-diver continues to fall at a constant speed. If he is moving slower than V_T, he accelerates. If he is moving faster, the increased drag slows him down. Thus, he approaches the terminal speed asymptotically. Typically, we may assume that he has reached his terminal value in about ten seconds.

The drag force depends on several factors, one being the cross-section area of the body. This is how a parachute works. A sky-diver face down with arms stretched out can reach about 200 km/h. Speed sky-divers take a head-down position and reach much higher speeds. Felix Baumgartner, jumping from 39 km, broke the speed record on 14 October 2012, reaching over 1340 km/h. This was at high altitude, where the air density is very small. Two years later, Alan Eustace broke Baumgartner's record when he jumped from a higher altitude of over 41 km.

EARTH'S SHAPE AND SPIN WON'T MAKE YOU THIN

Many of us struggle to lose weight, or at least to keep our weight within a manageable range. There is no easy way to do this, but could geophysics provide some assistance? The earth is approximately spherical, but there is a slight flattening towards the poles. This is a consequence of the rotation when the planet was forming during the early history of the solar system.

The gravitational attraction of a sphere of uniform density is the same as if all the mass were concentrated at the centre. Thus, it is the same everywhere on the surface. But is the earth a perfect sphere? Newton believed that rotation would bring about a flattening, with the polar radius smaller than that at the equator. This would make it an *oblate spheroid*, flattened like an orange. Dominique Cassini and his son Jacques took a contrary view: their measurements in France indicated that the earth is elongated at the poles, a shape called a *prolate spheroid*, more like a lemon.

If the earth were flattened, the length of a meridian arc of one degree of (geographic) latitude would increase from equator to poles. If the earth were elongated, as held by the Cassinis, the length of such a meridian arc would decrease towards the poles. But which was true? There was a lively controversy between the Orangemen and the Elder Lemons.

To resolve the issue, in 1735 the French Academy of Sciences proposed a geodetic expedition to take measurements of the meridian arc in polar and tropical regions. Two groups set out from Paris, one to Peru and one to Lapland. The northbound group, led by Pierre Louis Maupertuis, was back in Paris within a year or so.

It is an amazing fact that, by means of a humble pendulum, we can determine the geometric shape of the earth. We can calculate the relative strength of gravity at two latitudes by measuring the period of the pendulum at the two locations. In 1673, the French astronomer Jean Richer found that his pendulum clock, which kept good time in Paris, lost 2.5 minutes per day in Cayenne. This means that the gravitational attraction in Cayenne is about 0.35% *smaller* than in Paris. From this we can calculate that the effective force of gravity increases by about 0.5% as we go from equator to pole.

Among the scientists accompanying Maupertuis was Alexis Claude Clairaut, a notable French mathematician, geophysicist and astronomer. In his *Theory of the Figure of the Earth* (1743), Clairaut published a formula that allows the ellipticity of the earth to be calculated from surface measurements of gravity. An analysis of the Lapland measurements confirmed that the earth is an oblate spheroid, as Newton had predicted. On hearing of the result, Voltaire congratulated Maupertuis 'for flattening the earth and the Cassinis'.

Variations of gravity have important geophysical consequences, for inertial navigation, GPS location and so on. But, when it comes to weight control, they are no substitute for the old reliables, good diet and regular exercise. A man weighing 100 kg can lose about half a kilo by moving from Anchorage to Zanzibar but, of course, his body mass does not change, only the pull of the earth upon it.

THE TANGLED
TALE OF KNOTS ‖

We are all familiar with knots. Knots keep our boats securely moored and enable us to sail across the oceans. They also reduce the cables and wires behind our computers to a tangled mess. Many fabrics are just complicated knots of fibre and we know how they can unravel.

If the ends of a rope are free, any knot can be untied, but if the ends are spliced together the knot cannot be undone. To a mathematician, a knot is like a rope with the ends joined. It is intrinsically equivalent (or homeomorphic) to a circle. But extrinsically, knots differ in the way they are embedded in three-dimensional space.

Knot theory is a subfield of topology. A central problem of knot theory is to classify all possible knots and to determine whether or not two knots are equivalent. Equivalence means that one knot can be transformed into the other by a continuous distortion without breaking or passing through itself. Technically, this transformation is called an *ambient isotopy*.

In practice, knots are often distinguished by means of knot invariants. An *invariant* is a quantity that remains unchanged no matter how the knot is distorted. If its value differs for two knots, they cannot be equivalent. Many invariants are known, but no single one is sufficient to distinguish between all inequivalent knots.

Millions of knots and links have been tabulated since the start of the study of knot theory in the 1800s. The figure below shows a few elementary examples, a simple circular loop called the *unknot*, and a trefoil knot together with its mirror image.

Simple knots. On the left is the unknot. Centre and right are a trefoil knot and its mirror image.

All these are distinct: none of them can be transformed into another one. There are knot invariants, such as the Jones polynomial, that take different forms for each of these three knots.

In 1867 the physicist William Thomson, later Lord Kelvin, developed an atomic theory using knots. He speculated that atoms were vortices in the aether, different kinds of atoms being knotted in different ways. Each kind of atom would have a different kind of knot. Atoms could combine like tangled smoke rings.

This theory led nowhere in physics, but it inspired Kelvin's colleague Peter Guthrie Tait to create an extensive catalogue of knots, analogous to the periodic table of the elements. This triggered the mathematical study of knots as a sub-branch of topology. In some ways, Kelvin's idea was reminiscent of current efforts to model the fundamental forces of nature using string theory.

The unknot is similar to a circle, and we can regard it as the boundary of a disc-like surface. Surprisingly, we can find surfaces with knotted boundaries. In fact, every knot can be represented as the boundary of a surface: given any knot, we can construct an orientable (two-sided) surface having the knot as its boundary. In 1934 the German mathematician Herbert Seifert devised an algorithm for constructing such a surface for any knot.

Physical science often leads to the development of new fields of mathematics. Mathematicians may then carry the study of these fields far beyond the immediate needs of physics. This research may prove useful at a later stage in new problems in physics. This was the case with knot theory. Inspired by Kelvin's vortex atoms, it took on a life of its own, developed by mathematicians but ignored by physicists until recent decades. It is now important for the study of DNA entanglement and in unified theories of fundamental physics.

PLATEAU'S PROBLEM: SOAP BUBBLES AND SOAP FILMS

Bubbles floating in the air strive to achieve a spherical form. Large bubbles may oscillate widely about this ideal whereas small bubbles quickly achieve their equilibrium shape. The sphere is optimal: it encloses maximum volume for any surface of a given area. This was stated by Archimedes, but he did not have the mathematical techniques required to prove it. It was only in the late 1800s that a formal proof of optimality was completed by Hermann Schwarz.

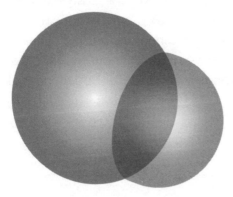

Computer-generated double bubble.

The famous problem known as Plateau's Problem is this:

> *Given a simple closed curve C in three dimensions, prove that there is a surface of minimum area having C as its boundary.*

Mathematically, the problem falls within the ambit of the calculus of variations.

Plateau formulated a set of empirical rules, now known as Plateau's Laws, for the formation of soap films.

1. Soap films are made of components that are smooth surfaces.

2. The mean curvature of each component of a soap film is constant.

3. Soap films connect in threes, along curves meeting at angles arccos $(-1/2) = 120°$.

4. These curves meet in groups of four, at vertices with angles arccos $(-1/3) \approx 109.5°$.

These laws hold for surfaces of minimal area, subject to constraints such as fixed enclosed volumes or specified boundaries.

Soap films are tense: they act like a stretched elastic skin, trying to attain the smallest possible area. The curvature of a soap film is related to the pressure difference across the surface by a nonlinear partial differential equation called the Young–Laplace equation. Surface energy is proportional to area, and surface tension acts to minimise the area. Thus, soap films assume a minimal area configuration.

If the pressure is equal on both sides, then there is no force normal to the surface. Therefore, the mean curvature is also zero. If the pressure differs on the two sides of a soap film, there is a force normal to the surface, which must therefore be curved so that the surface tension can balance the pressure force. Examples of surfaces with zero mean curvature are the catenoid (generated by rotating a catenary about its directrix) and the helicoid (generated by the lines from a helix to its axis).

Catenoid (left) and helicoid (right).

Plateau's Problem was solved under certain conditions by the American mathematician Jesse Douglas in 1931, and independently at about the same time by Tibor Radó. In 1936, Douglas was awarded the first Fields Medal for his work. Recently, Jenny Harrison of the University of California Berkeley published a general proof of Plateau's Problem, including non-orientable surfaces and surfaces with multiple junctions.

DOUBLE BUBBLE CONJECTURE

Plateau's original problem dealt with surfaces of minimum area without pressure differences and with boundaries that are simple closed curves. When we consider bubbles, an additional complication arises since there are variations in pressure, and the surfaces contain multiple components connected in complicated ways.

With just a single component, the solution is a spherical bubble. For a *double bubble*, there are two spherical outer surfaces joined by a spherical inner interface. A cross-section of a double bubble is shown below. For a static solution all forces must be in balance. At point P, three forces of equal magnitude in three different directions must sum to zero. The only way in which this can happen is if they meet at angles of 120°.

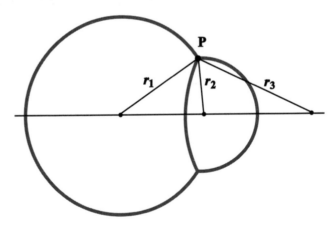

The acute angles at point P are each 60°. Using the sine rule, it is easy to show that

$$1/r_1 + 1/r_3 = 1/r_2$$

This result also follows from a physical argument: for balance of the forces at the interior interface, the pressure difference $p_1 - p_2$ must equal the surface tension, proportional to the curvature $1/r_2$. But p_1 is proportional to $1/r_1$ and p_2 is proportional to $1/r_3$, which leads to the desired result.

The double bubble comprising three spherical caps is the minimal surface enclosing two fixed volumes of air. Surprisingly, this conjecture was not proved until as recently as 2002 (Hutchings *et al.*, 2002).

SOURCES

Hutchings, Michael, Frank Morgan, Manuel Ritoré and Antonio Ros (2002), 'Proof of the double bubble conjecture', *Annals of Mathematics*, 2nd Series 155 (2), 459–89.

THE STEINER MINIMAL TREE PROBLEM

Steiner's minimal tree problem is this: *Find the shortest possible network interconnecting a set of points in the Euclidean plane*. If the points are linked directly to each other by straight line segments, we obtain the *minimal spanning tree*. But Steiner's problem allows for additional points – now called Steiner points – to be added to the network, yielding *Steiner's minimal tree*. This generally results in a reduction of the overall length of the network.

Jakob Steiner (1796–1863), a Swiss mathematician, was one of the greatest geometers of all time. Born near Berne, he moved to Germany in 1818, and spent most of his life in Berlin, where he was on friendly terms with Jacobi and Abel. The problem we now call Steiner's problem was originally posed in purely geometric terms, but its solution involves mathematical techniques from combinatorics, analysis and computational science. The problem is relevant for a wide range of applications, including communications, power grids, transport networks, electric circuit layout, facility location, pipeline networks and the efficient design of microchips.

SOLUTIONS FOR FOUR POINTS IN A RECTANGLE

The book *What is Mathematics?* by Richard Courant and Herbert Robbins, which first appeared in 1941, was very influential. Written in an expository way and accessible to a wide audience, it continues to be popular and is still in print. In Chapter VII, 'Maxima and Minima', Courant and Robbins discuss the Steiner problem at length. Among other things, they point out that the solution is not necessarily unique. For example, if four points are given at the corners of a square, there are two solutions, as shown here:

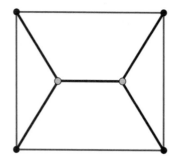 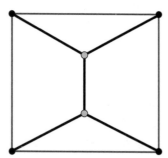

For a unit square, each of these networks is of length 1 + √3. It is interesting that neither solution has the same degree of symmetry as the square: the solutions have less symmetry than the problem itself!

Courant and Robbins also discussed an 'analogue method' of solving the Steiner problem: an empirical solution can be found using soap films. If two sheets of glass joined by perpendicular struts are dipped in a soap solution, a film forms comprising plane sections meeting at angles of 120°. Soap films provide a local minimum configuration, but not always the global minimum.

Let us start with the solution for a rectangle with large breadth and small height (panel 1 in the figure below). Imagine a soap film in the form of the minimum Steiner tree. Now we distort the rectangle, increasing the height but keeping the length of the perimeter constant. The topological form of the solution (with a horizontal strut) remains unchanged as it passes through a square shape (panel 3). At some stage (panel 4) the Steiner points coalesce and the solution forms the two diagonals. This is an unstable configuration and immediately jumps to the form in panel 5, with a vertical strut: the topology has changed. It retains this form as the height is further increased.

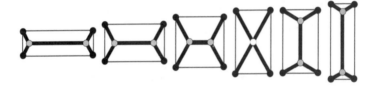

Now we reverse the process (see figure below): as the height is gradually decreased, the form of the soap film changes, but it maintains its topology until the shape passes beyond a square shape (panel 3). When the Steiner points coalesce (panel 4), the unstable solution jumps to a different topology, with a horizontal strut, and retains this form as the rectangle flattens further.

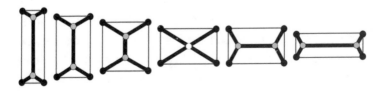

We notice that the solution assumed for four points in a square is dependent on the route through which the solution is reached. There is a hysteresis effect whereby, when the corner points move through a sequence of shapes and return to their original position, the final solution may be different from the initial one.

COMPUTATIONAL SOLUTIONS

In 1968 Edgar Gilbert and Henry Pollak of Bell Laboratories published a comprehensive study, 'Steiner Minimal Trees'. This set the scene for the development of fast numerical algorithms for solving the problem. According to Brazil *et al.* (2014), this paper 'helped to kindle a fascination with ... [the Steiner] problem in the mathematical community that has lasted to the present day'.

In a few special cases, Steiner's problem can be solved exactly. For a regular polygon with $n > = 6$ sides, the polygon itself (less one edge) is the solution: the Steiner minimal tree coincides with the minimal spanning tree; no additional Steiner points are required. However, in general, we cannot obtain an analytical expression for the solution and must use numerical methods to find it.

For large n, the Steiner tree problem is computationally expensive (technically, it is NP-hard). To solve it, 'heuristic methods' are used. These methods yield a solution that may not be optimal, but is close enough to be useful. Normally, they provide a solution that is within a few per cent of the minimum solution. With heuristics, solution of the Steiner problem is computationally feasible for thousands of terminal points.

There is a variant of the Steiner problem, called the rectilinear Steiner problem, in which all the line segments must be either horizontal or vertical lines. This problem uses the so-called taxicab metric and is relevant for design of VLSI electric circuits.

SOURCES

Brazil, M., R. L. Graham, D. A. Thomas and M. Zachariasen (2014), 'On the history of the Euclidean Steiner tree problem', *Archive for History of Exact Sciences*, **68**(3), 327–54.

WHO WANTS TO BE A MILLIONAIRE?

Topology is the mathematics of shapes. Geometers consider lengths, angles and other details. Topologists are more interested in overall, global aspects, properties that survive when a shape is distorted or stretched, as long as it is not torn. For a topologist, a sphere is the same as a cube: they are both three-dimensional shapes without any holes.

A doughnut has a hole, so it is different from a sphere. Spectacle frames have two holes, as have handcuffs; and so on. The number of holes determines the topological character of a surface. Two-dimensional surfaces have been completely classified using this idea.

The French mathematician Henri Poincaré knew that a sphere is characterised by the property of having no holes. In 1904 he posed the corresponding question for hyperspheres. The hypersphere is the set of points that are equidistant from a central point in four-dimensional space. It is much more difficult to visualise than the usual sphere.

How did Poincaré make this precise? If you draw a loop on a sphere (say, the equator) you can shrink it to a point by moving it around. But a loop around a doughnut gets blocked by the hole and cannot be shrunk to a point. If every loop can be contracted to a point, there are no holes.

The question that Poincaré posed is whether the hypersphere is the only solid shape with no holes. But he could not answer the question. It was considered to be one of the most important outstanding problems in mathematics. Numerous claims of a proof were made but, until recently, all were shown to be flawed. The problem seemed unassailable.

In 1982, the American mathematician Richard Hamilton came up with an ingenious plan of attack. He would treat curvature as something fluid that could change over time. Heat always flows from hotter to colder places, smoothing out differences. What if we tried to smooth out a shape in this way?

Hamilton's idea was to start with an arbitrary shape without holes, and simplify it by smoothing out the curvature, allowing it to flow around the shape. He was hoping that this would always lead to a hypersphere, proving Poincaré's conjecture. Sadly, there was a major hitch: the flow sometimes caused the curvature to grow without limit, forming a *singularity*.

In 2002 and 2003 the Russian mathematician Grigori (Grisha) Perelman posted three preprints on the internet. He used Hamilton's strategy but overcame the difficulties with singularities by performing 'surgery' on the shape. He found that every shape without holes would eventually become a hypersphere.

Perelman's preprints left several gaps and it was not clear whether or not he had proved Poincaré's conjecture. But his work was intensively studied by several high-profile mathematicians, who confirmed his result.

In 2006, Perelman was offered a Fields Medal – comparable in prestige to a Nobel Prize – for his work. He refused the prize, saying that he was not interested in money or fame. In 2010 the Clay Mathematics Institute announced the award of the first Millennium Prize to Perelman. This would have made him a millionaire overnight. But Perelman turned down the prize, saying that his contribution to proving the conjecture was no greater than Hamilton's.

Perelman now lives reclusively in St Petersburg and it is unclear whether he has abandoned mathematics. Let us hope not, for he is fearsomely brilliant. Such people are very rare, and six Millennium Prize Problems remain to be solved.

A loop on a 2-sphere can be continuously shrunk to a point, and the surface is simply connected. The Poincaré conjecture states that every simply connected, closed 3D manifold is homeomorphic to a 3-sphere.

THE KLEIN 4-GROUP ||

What is the common factor linking book-flips, solitaire, twelve-tone music and the solution of quartic equations? Answer: K_4.

Take a book, place it on the table and draw a rectangle around it. How many ways can the book fit into the rectangle? Clearly, once any single corner of the book is put at the top left corner of the rectangle, there is no further leeway; the positions of the remaining three corners are determined. Thus, there are four ways the book can fit into the rectangle. They are shown in this figure.

The four symmetric configurations of a book under 3D rotations.

The four orientations of the book can be described in terms of simple rotations, starting from the upright configuration:

- Place the book upright with front cover up (that is, do not move it).

- Rotate through 180° about X-axis (horizontal line through centre).

- Rotate through 180° about Y-axis (vertical line through centre).

- Rotate through 180° about Z-axis (line through centre perpendicular to book).

We use the four symbols {I, X, Y, Z} for these operations. We denote the combination or composition of two operations by *. It is clear that any operation performed twice brings us back to the original position:

$$X * X = Y * Y = Z * Z = I$$

Moreover, combining any two of {X, Y, Z} gives the third, for example, X * Y = Z. This means that X * Y * Z = I. Drawing up a full table of combinations of two operations, we get the following table (the Cayley table):

*	I	X	Y	Z
I	I	X	Y	Z
X	X	I	Z	Y
Y	Y	Z	I	X
Z	Z	Y	X	I

This is a simple example of a group, a set of elements together with a rule for combining them. A group operation must satisfy four conditions: closure, associativity, identity and inverse. The book symmetries are a realisation of the Klein 4-group, K_4.

THE KLEIN 4-GROUP

The group K_4 was introduced by Felix Klein in his study of the roots of polynomial equations, solution of cubics and quartics and the unsolvability of the quintic equation. The orientations of a book, or symmetries of a rectangle, are just one way to describe the group. A more formal representation is to express the rotation operations {I, X, Y, Z} as three-by-three matrices. The identity operation and the rotations through 180° about the axes are

$$I = \begin{pmatrix} 1 & 0 & 0 \\ 0 & 1 & 0 \\ 0 & 0 & 1 \end{pmatrix} \quad X = \begin{pmatrix} +1 & 0 & 0 \\ 0 & -1 & 0 \\ 0 & 0 & -1 \end{pmatrix} \quad Y = \begin{pmatrix} -1 & 0 & 0 \\ 0 & +1 & 0 \\ 0 & 0 & -1 \end{pmatrix} \quad Z = \begin{pmatrix} -1 & 0 & 0 \\ 0 & -1 & 0 \\ 0 & 0 & +1 \end{pmatrix}$$

It is a simple matter to show that these matrices satisfy the same multiplication table as shown above. Thus {I, X, Y, Z} is a representation of the Klein 4-group in the group of special orthogonal transformations SO(3) in R^3.

The book – or at least its infinitely thin idealisation – can be thought of as a two-dimensional object. Suppose we wish to remain in the plane; then only the operations I and Z can be performed. But we can replace the other two rotations by reflections in the X and Y axes. Then we represent the four operations by four two-by-two matrices operating in the X–Y plane.

$$I_2 = \begin{pmatrix} 1 & 0 \\ 0 & 1 \end{pmatrix} \quad X_2 = \begin{pmatrix} +1 & 0 \\ 0 & -1 \end{pmatrix} \quad Y_2 = \begin{pmatrix} -1 & 0 \\ 0 & +1 \end{pmatrix} \quad Z_2 = \begin{pmatrix} -1 & 0 \\ 0 & -1 \end{pmatrix}$$

Again, it is simple to show that the matrices $\{I_2, X_2, Y_2, Z_2\}$ have the same Cayley table as K4. Thus, they are a representation of the Klein 4-group in terms of matrices in the orthogonal group O(2) of isometries in two-dimensional space R^2. The configurations are shown below.

The four symmetric configurations of a book under two-dimensional reflections and rotations.

OTHER APPLICATIONS

The Klein 4-group is also useful for musicians working on twelve-tone composition. In the twelve-tone technique – also called dodecaphony – the composer starts with a tone row containing all the notes of the chromatic scale. This tone row can then be transformed using reflection (left–right flip), inversion (up–down flip) or a combination of these (rotation through 180°). These transformations are completely equivalent to the symmetries of a rectangle, embodied in the group K_4.

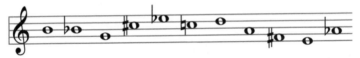

A *prime tone row. Transformations are the retrograde, inversion and retrograde-inversion.*

Finally, we mention that the Klein 4-group occurs in Galois' Theorem, which gives conditions under which a polynomial equation is soluble in radicals: K_4 is a component in the chain of subgroups $S_4 > A_4 > K_4 > I$, giving a basis for the solution of the quartic equation.

TRACING OUR MATHEMATICAL ANCESTRY: THE MATHEMATICS GENEALOGY PROJECT

There is great public interest in genealogy. Many of us live in hope of identifying some illustrious forebear, or enjoy the frisson of having a notorious murderer in our family tree. Numerous websites facilitate the search and online census returns have led to an explosion of interest in the field.

Academic lineages can also be traced. For example, the Mathematics Genealogy Project (MGP) has assembled a huge database of mathematicians. Parent–child relationships are replaced by links between students and their doctoral supervisors or, before PhDs were common, links between mentors and protégés. Usually this means one-parent families, but there are often two or more advisers, making the structure or topology of the family tree more complex and more interesting.

MGP was set up in 1996 by Harry Coonce, then at Minnesota State University. It was a labour of love, and he worked incessantly on developing the project, with only meagre resources. It is now hosted by North Dakota State University (www.genealogy.math.ndsu.nodak.edu). The number of entries has grown steadily, with about a thousand new ones every month, and is now at about 200,000. The term 'mathematician' is interpreted quite broadly and includes computer scientists and theoretical physicists.

To give a concrete example, click back through the links from Grigori Perelman, the mathematician who in 2003 proved the Poincaré conjecture, and you get to Andrei Markov and Pafnuty Chebyshev, giants of probability and statistics. Before them comes Nikolai Lobachevsky, famous for inventing hyperbolic geometry. Erasmus is also among Perelman's 'ancestors'. The trail eventually goes cold, but not until the beginning of the fourteenth century.

Generally, the links stop somewhere like medieval Oxford or Renaissance Bologna. Since many mathematicians have more than one adviser, the number of forebears grows substantially as we go further back, and the chances of having a famous ancestor are high. Still, it is great fun to find that Newton, Euler or Gauss is in one's academic family tree, even if one in three mathematicians can make a similar claim. And if you get to Newton, you go back to Galileo too!

Mentor–mentee links are not guaranteed to provide a meaningful chain between modern mathematicians and the 'greats' of earlier times. But as long as they are not over-interpreted, they do provide valuable information about the connections between mathematicians of different eras and locations. Assuming that strong students identify good supervisors and excellent advisers take only the best candidates, then, by and large, better students should link back to more famous mathematicians. A sparkling academic lineage may be only loosely coupled to academic prowess but – to be honest – we all get a thrill to link to someone famous, staking a claim to the glittering legacy of mathematics.

The Mathematics Genealogy Project database is like a giant genealogical chart, with the interconnections forming a network or graph. Currently, entries go back to about AD 1300. The ultimate aim is to include all mathematicians in the world. Who would not love to be a descendant of Archimedes? Wouldn't it be wonderful if we could establish a link back to Ancient Greece, and to the mathematicians of Babylonia, India and China? But that is a dream.

CAFÉ MATHEMATICS IN LVOV ‖

For 150 years the city of Lvov was part of the Austro-Hungarian Empire. After Polish independence following World War I, research blossomed and between 1920 and 1940 a sparkling constellation of mathematicians flourished in Lvov.

Zygmunt Janiszewski, who had been awarded a doctorate by the Sorbonne in 1911, had a vision of Polish mathematical greatness and devised a programme for its achievement. He advocated that Polish mathematicians should specialise in a few clearly defined fields. This would ensure common interests and foster a culture of collaboration.

A plan like this can succeed only if there are talented mathematicians to carry it out. Fortunately, while there was no strong tradition of excellence, several brilliant Polish mathematicians emerged around that time. The leading lights were Hugo Steinhaus, with a doctorate from Göttingen – then the Mecca of mathematics – and

Stefan Banach, who would become the greatest Polish mathematician.

Diverse contributions to mathematics were made by the Lvov School, earning it worldwide admiration. Names like Banach, Sierpiński, Kac and Ulam occur frequently in modern textbooks. They were concerned with fundamental aspects of mathematics: axiomatic foundations of set theory, functions of a real variable, the nature of general function spaces and the concept of measure.

The year 1932 saw the publication of Banach's monograph on normed linear spaces. It contained many powerful results. His genius was to combine different fields of mathematics. He treated functions as points in an abstract space that was linear, with a concept of distance and an absence of 'gaps': a complete, normed linear space, a fusion of algebra, analysis and topology. It proved eminently suitable for the development of the field called functional analysis,

Banach's monograph had a major influence and his notation and terminology were widely adopted. His spaces quickly became known as Banach spaces and they have played a central role in functional analysis ever since. They also served as a foundation for quantum mechanics. Banach's monograph established the international importance of the Lvov School.

Many of the mathematical breakthroughs in Lvov resulted from collaborations, and the majority of the publications are the work of two or more authors. The mathematicians used to meet regularly in cafés, discussing mathematics late into the night. Their favourite haunt was the Scottish Café, undoubtedly the most mathematically productive café of all time. The table tops were of white marble on which mathematics could be easily written (and erased).

Later, Banach's wife bought a large notebook for the group – the famous 'Scottish Book' – in which problems and solutions were recorded. This was kept in the café and was available to any mathematicians who visited. Ultimately, it contained about 200 problems, many of which remain open to this day. One problem caused a media sensation when it was finally solved. Stanisław Mazur had offered a live goose for a solution and in 1973 the Swedish mathematician Per Enflo travelled to Warsaw to collect his prize from Mazur for solving 'the goose problem'.

The Scottish Café exemplified the synergy and camaraderie that pervaded Polish mathematics in the inter-war years. World War II changed everything. Polish culture was systematically eradicated. Steinhaus managed to escape execution by assuming a false identity. Banach survived to witness the defeat of Nazism but died shortly afterwards. Today, the city is Lviv, a major centre of culture in western Ukraine. The Golden Age of the Lvov School has passed into history.

THE KING OF INFINITE SPACE: EUCLID AND HIS ELEMENTS

The Elements – far and away the most successful textbook ever written – is not just a great mathematics book. It is a great book. There is nothing personal in the book, nothing to give any clue as to the author. Yet virtually everything of importance in classical Greek mathematics is contained in it.

Almost nothing is known of the life of Euclid. He flourished in Alexandria around 300 BC. He may have taught at the great Library of Alexandria, founded by Ptolemy I. He may have attended Plato's Academy. Plato (428–347 BC) was devoted to geometry and, allegedly, the inscription over the entrance to his Academy read 'Let none but Geometers enter here'.

Euclid came perhaps fifty years after Aristotle (384–322 BC) and was certainly familiar with Aristotle's logic. Euclid organised and systematised the work of earlier geometers, also adding substantially to their work. His system, which was truly innovative, relied on basic assumptions, called axioms and common notions. Most of these might be considered beyond dispute, for example 'if equals are added to equals, the wholes are equal'. Who could argue with this? Today, we might write 'If A = B and C = D, then A + C = B + D'.

Euclid imposed an order on mathematics, creating an axiomatic system that endures to this day. His aim was to deduce a large number of theorems or propositions from a small number of assumptions or axioms. Indeed, he proved nearly five hundred theorems from just five axioms. The deductions had to be logically unimpeachable, with each step following clearly from the one before.

The organisation and structure of *The Elements* is brilliant. It begins with definitions: 'A point is that which has no parts'; 'A line is length without breadth'. There are in all 23 definitions. Then come five axioms and five 'common notions'. The axioms, or postulates, are specific assumptions that may be considered as evidently true.

Two of the axioms postulate that it is possible to draw a unique straight line between two points, and to describe a circle with a given centre and radius. The idea of using a straight edge and compass is implicit in these axioms; Euclid does not actually mention these instruments in the axioms or elsewhere in *The Elements*, but they are essential to his geometry. The common notions are more general than the axioms, and are suppositions 'on which all men base their proofs'.

The Elements is arranged in thirteen books, in which nearly 500 (467, to be precise) propositions or theorems are proved. Book I has 48 propositions or theorems. The first concerns the construction of an equilateral triangle, a three-sided figure, all sides having equal length. Theorem 4 states that two triangles having two sides and the angle formed by them equal are congruent, or equal in all respects. Euclid's proof involves the movement of triangles, something not justified by him. Indeed, this is just one of many shortcomings identified by modern commentators.

The tension builds up through Book I from one theorem to the next, leading to an impressive climax, the proof of the Pythagorean Theorem and its inverse in Theorems 47 and 48. Known to the Babylonians and rediscovered by Pythagoras around 500 BC, it is rigorously proved here, perhaps for the first time, by Euclid.

The final book, Book XIII, presents the theory of the three-dimensional solids known as the platonic solids. Each of these has faces that are regular polygons, with all faces identical and all angles between adjacent faces equal. The best known is the cube, with six similar square faces. There are no more than five such solids, and this was known to the classical Greek mathematicians.

There are schematic illustrations accompanying every theorem. These are an ingenious means of clarifying the mathematical proofs. What were Euclid's original diagrams like? Who drew them? We have no idea. The trail back to the original Greek sources goes cold around AD 850. No original writings of Euclid's *Elements* remain, just copies of copies, some translated from Greek to Arabic to Latin and back to Greek again.

During the Renaissance, the Western world learned of Euclid from Arabic translations of the Greek text, further translated into Latin. The earliest surviving version of the Greek text is the 'Vatican Euclid' (Vat. graec. 190) dating from the ninth century. The oldest surviving translation into Latin was by Adelard of Bath (c. 1080–1152), an English natural philosopher.

THE PARALLEL POSTULATE

For many centuries, geometers worried about Euclid's fifth postulate, which, in one form, states that through any point not on a line one and only one line parallel to the first can be drawn. This seems less obvious and less natural than the other four axioms, and great efforts were made to derive it as a theorem, all being unsuccessful. In 1733 the Italian Jesuit mathematician Gerolomo Saccheri published a treatise called '*Euclides ab omni naevo vindicatus*' ('Euclid freed of every flaw'). Since he was unable to deduce the parallel postulate from the other four axioms, Saccheri assumed that it was false, seeking a contradiction. But none was forthcoming, and Saccheri came very close to discovering non-Euclidean geometry.

While still in his teens, Gauss followed a similar line to Saccheri, but pursued it further to find a self-consistent system in which the parallel postulate is negated. However, Gauss did not publish this work, either at the time or later in his life. But when Bolyai and Lobachewski produced their results, Gauss revealed that, for him, there was nothing new. This was not the only occasion on which Gauss behaved in this infuriating manner.

The existence of several geometries was disquieting for nineteenth-century mathematicians. They might well have asked: 'Which geometry is right? Which one applies to the real world?' Felix Klein addressed the problem in 1872 in his inaugural discourse at the University of Erlangen. His lecture was effectively a manifesto and it had a profound influence on subsequent research, becoming known as the Erlangen Program.

Klein sought a unifying principle for the various versions of geometry and recognised that the essential character of each was embodied in the set of transformations that were permitted. In each case, these transformations formed a group. For example, Euclid proved his Theorem 4 by moving a triangle. Euclidean geometry allows the translation of a triangle from one place to another, its rotation about a point or its reflection in a line. All these transformations preserve the distance between points. They are called isometries and comprise what we call the Euclidean group.

The Elements was central to intellectual life for more than two millennia. Anyone aspiring to an education had at least some familiarity with Euclid. During the past century, classical geometry has fallen out of fashion and out of favour. Yet Euclid's axiomatic method, in which theorems are deduced logically from a small number of assumptions, remains at the core of mathematics. The spirit of Euclid continues to inspire us and to guide progress in mathematics.

(Berlinski, David, 2013: *The King of Infinite Space: Euclid and His Elements*. Basic Books.)

GOLDEN MOMENTS ‖

Suppose a circle is divided by two radii and the two arcs a and b are in the golden ratio:

$$b / a = (a + b) / b = \varphi \approx 1.618$$

Then the smaller angle formed by the radii is called the golden angle. It is equal to about 137.5° or 2.4 radians. We will denote the golden angle by γ. Its exact value as a fraction of a complete circle is $(3 - \sqrt{5}) / 2 \approx 0.382$ cycles.

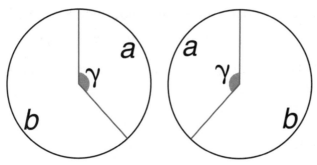

The golden angle is found in many contexts in nature. In phyllotaxis, it is the angle separating successive leaves, petals or florets in many flowers.

The golden rectangle is a rectangle whose width and height are in the ratio of the golden number $\varphi \approx 1.618$. Allegedly, it has great aesthetic appeal, and a great deal – of both sense and nonsense – has been written about the occurrence of the golden number in art.

As the hands of a clock turn, the angle between the minute and hour hands varies continuously from zero to 180° and back to zero in a period slightly longer than an hour. More precisely, the hands come together every 12/11 hours. It is convenient to measure angles in cycles, or units of 360°, and time in hours and decimal fractions of an hour. Then the angular speed of the minute hand is 1 cph (cycle per hour) and that of the hour hand is 1/12 cph. So the relative speed of the hands is $(1 - 1/12) = 11/12$ cph. Thus, the minute hand catches up to, or laps, the hour hand after a period of 12/11 hours.

We define *moments of synchrony* to be those times when the two hands overlap. Clearly, midnight and midday are two such moments. There are eleven moments of synchrony in each twelve-hour period. They occur at times:

$$T_S(N) = 12(N-1)/11, \quad N = 1, 2, 3, \dots, 11$$

So $T_S(1)$ is at midnight or midday, $T_S(2)$ is at 12/11 hr or 1h 5m 27.27s and so on.

GOLDEN MOMENTS

Between every two moments of synchrony there are two *golden moments*, that is, times when the angle between the hands of the clock is equal to the golden angle γ.

Since the relative angular speed is 11/12 cph, the time after synchrony of the next golden moment occurs after a time *t* such that

$$(11 / 12)\, t = \gamma \quad \text{or} \quad t = (12 / 11)\, \gamma$$

This implies an interval of 0.416727 hours or almost exactly 25 minutes. The next golden moment is when

$$(11 / 12)\, t = 1 - \gamma \quad \text{or} \quad t = (12 / 11)\, 0.618 = 0.674$$

which is at about 40m 27s after synchrony. Further pairs of golden moments occur at intervals of about one hour five and a half minutes.

The first two golden moments after midnight. There are 44 such moments every day.

In total, there are 44 golden moments in a day. Do they have any aesthetic or psychological significant? Presumably not, but they are mathematically interesting.

Exercise: You may wish to construct a complete table of golden times so that you can savour the moments.

Note: The clock faces are drawn using Mathematica code downloaded from the blog of Christopher Carlson, at http://blog.wolfram.com/2007/07/09/always-the-right-time-for-mathematica/

MODE-S EHS: A NOVEL SOURCE OF WEATHER DATA

It has often happened that an instrument designed for one purpose has proved invaluable for another. Galileo observed the regular swinging of a pendulum. Christiaan Huygens derived a mathematical formula for its period of oscillation and used a pendulum to develop the first precision timepiece. Later, a pendulum was used by Maupertuis to measure the strength of the earth's gravitational attraction and thus determine the shape of the earth.

A recent example of unexpected utility is the meteorological application of transponders developed for air traffic management and safety. Accurate wind information for the upper atmosphere is a key requirement for weather prediction. Currently, most wind data comes from weather balloons, wind profilers, Doppler radars and satellites. Mode-S EHS, a novel source of wind data from aircraft flight levels, is now helping us to make more accurate weather forecasts.

Under European regulations, all large aircraft must carry Mode-S EHS enhanced surveillance navigation apparatus. Aircraft equipped with EHS transponders are interrogated every four seconds by ground-based radar and, in response, send information on aircraft position, flight level, magnetic heading, air speed and ground speed. Air traffic control (ATC) monitors this data to ensure efficient and safe operations.

How does this yield wind data? Suppose a plane is heading eastward at 200 metres per second. Its position is known accurately by the satellite-based global positioning system or GPS. Four seconds later, it should be 800 metres east of its initial position. But suppose there is a wind of 50 m/s from the south-east. This will slow the plane and cause it to drift to the north.

The GPS location shows precisely where the plane has gone in four seconds and determines the ground speed. Since ground speed **G** is the vector sum of air speed **A** and wind speed **W**, a simple vector calculation gives us the wind. The figure illustrates the three vectors and shows how the wind **W** is given by **G** – **A**. What a delightfully simple application of vectors; what a shame that vectors have been dropped from Leaving Certificate maths!

In practice, further adjustments are required to produce high-quality wind estimates. With these corrections and calibrations, Mode-S EHS winds are of an accuracy comparable to conventional wind observations. Since air traffic for Europe is planned and co-ordinated by Eurocontrol, the novel wind vectors are potentially available throughout the European region. Currently, the coverage includes UK, Benelux and Germany, and more countries are expected to implement EHS radar systems soon. Met Éireann acquires Mode-S data using a receiver in Valentia Observatory and is exploring its value in the Harmonie forecasting model.

Siebren de Haan, a senior scientist at the Royal Netherlands Meteorological Institute, has examined the impact of the new observations on forecast accuracy and has found that Mode-S EHS observations are beneficial for short-range wind forecasts. This facilitates air traffic management and leads to greater fuel efficiency.

In addition to wind vectors, the Mode-S EHS data can be used to deduce air temperature data. Temperature is proportional to the squared ratio of the air speed and the Mach number, the speed relative to the local speed of sound. Thus, a simple calculation gives the temperature in terms of measured quantities. But that is another story.

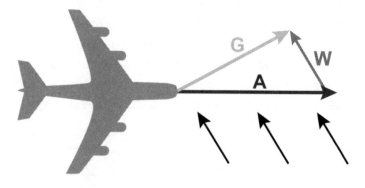

The air speed is A, the wind speed is W and the ground speed is G. Since the ground speed is the resultant (vector sum) of air speed and wind speed, a simple vector subtraction gives the wind speed: W = G – A.

FOR GOOD COMMUNICATIONS, LEAKY CABLES ARE BEST

In Wicklow town an obelisk commemorates Robert Halpin, a master mariner born at the nearby Bridge Tavern. Halpin, one of the more important mariners of the nineteenth century, 'helped to make the world a global village' by connecting continents with submarine telegraph cables.

Halpin left home aged just 11 and had many nautical adventures before taking command of the leviathan SS *Great Eastern*, a brainchild of Isambard Kingdom Brunel. Its mission was to lay a telegraph cable across the Atlantic Ocean, from Valentia Island to Newfoundland, a distance of some 2600 nautical miles. The consulting engineer was William Thomson, later Lord Kelvin, the Belfast-born scientist who oversaw the laying of the cables.

Telegraph speed on the first cable, completed in 1858, was very slow, and it failed within a week. A second cable was more robust and faster, carrying messages at eight words a minute – about one Morse character per second. Thomson had analysed the transmission properties of cables and had argued that electrical resistance and leakage, or conduction through the imperfect insulation, should be minimised for best results.

As signals travel along a telegraph cable, they lose energy and also suffer distortion. Different signal components travel at different speeds, causing corruption of the message. To ensure distortion-free transmission, components of different frequencies must travel at the same speed. Early cables were designed to reduce energy loss, with no effort to avoid distortion.

Thomson had used an equation of parabolic type, with solutions that diffuse and dampen out but that are not wave-like. However, he omitted a crucial factor: when a current flows through a wire, it generates a magnetic field, which acts back on the current, inhibiting the flow. This *self-inductance* has a major impact on the transmission characteristics of a telegraph cable.

The self-educated and eccentric English electrical engineer Oliver Heaviside derived a mathematical equation – called the telegraph equation – which included inductance. This effect completely changes the nature of the solutions so that they have a wave-like character. Mathematicians call equations of this type hyperbolic equations.

Heaviside's idea was simple: a signal that has been damped can easily be amplified, but one that is distorted is difficult, if not impossible, to recover. So the design should focus on removing distortion. He showed that, to achieve distortion-free transmission, some electrical leakage between the conductors of a cable is essential.

Heaviside found that two quantities, one involving resistance and one involving leakage, must have their arithmetic and geometric means equal. But, as we learn in school, this happens only if the quantities themselves are equal. This gave him a condition on the properties of the cable that are needed to avoid distortion of the signal.

The upshot of Heaviside's mathematical analysis was that cable designers should not try to reduce leakage but, rather, should increase the inductance to achieve the balance required for distortion-free transmission. Heaviside was a controversial character, in regular conflict with the scientific establishment, and the engineers responsible for cable construction were not inclined to listen to him. They either did not understand or did not believe his theory.

As a result, it was several decades before Heaviside's brilliant ideas were implemented. But from the beginning of the twentieth century, distortion-free telegraphy was widely implemented and became an industry standard. Transmission speeds could then exceed 100 words per minute.

TAP-TAP-TAP
THE COSINE
BUTTON

Tap any number into your calculator. Yes, any number at all, plus or minus, big or small. Now tap the cosine button. You will get a number in the range [-1,+1]. Now tap 'cos' again and again, and keep tapping it repeatedly (make sure that angles are set to radians and not degrees). The result is a sequence of numbers that converge towards the value

$$0.739085 \ldots$$

What you are calculating is the iterated cosine function

$$\cos(\cos(\cos(\ldots \cos(x0)))).$$

where x_0 is the starting value. As the process converges, you have found the solution of the equation

$$\cos(x) = x$$

This is called a fixed point of the cosine mapping. If we plot the two functions $y = x$ and $y = \cos(x)$, we see that they intersect in just one point, $x = 0.739085 \ldots$.

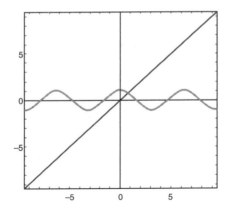

COBWEB PLOT

We can represent the iterative process of repeating cosine calculations using a diagram called a cobweb plot. On this visual plot, a stable fixed point gives rise to an inward spiral. For a function $f(x)$, we first plot the diagonal $y = x$ and the curve $y = f(x)$. We are looking for a point where the two graphs cross.

Let us start with a point on the diagonal, (x_0, x_0) on the diagonal. Moving vertically to the curve $y = f(x)$, we get the point $(x_0, f(x_0))$. Now move horizontally to the diagonal again, to $(f(x_0), f(x_0))$. Moving vertically and horizontally again between the curves, we get $(f(x_0), f(f((x_0)))$ and $(f(f(x_0)), f(f(x_0)))$. By means of these alternately horizontal and vertical moves, we generate the sequence

$$\{ x_0, f(x_0), f(f(x_0)), f(f(f(x_0))), \ldots \}$$

Under favourable circumstances, the sequence converges to the fixed point of the mapping $y = f(x)$. The process is illustrated in the following diagram.

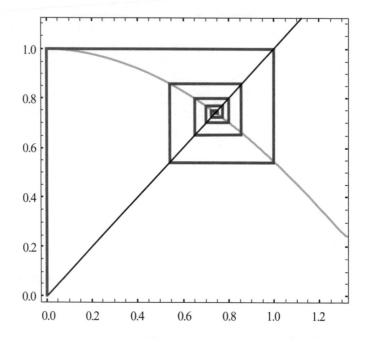

Fixed point theorems (FPTs) give conditions under which a function f(x) has a point such that f(x) = x. FPTs are useful in many branches of mathematics. Among the most important examples is Brouwer's FPT. This states that for any continuous function $f(x)$ mapping a compact convex set into itself, there is a point x_0 such that $f(x_0) = x_0$. The simplest example is for a continuous function from a closed interval I on the real line to itself.

More generally, Brouwer's theorem holds for continuous functions from a convex compact subset K of Euclidean space to itself. Another FPT is that of the Polish mathematician Stefan Banach. In technical terms, this states that a contraction mapping on a complete metric space has a unique fixed point.

THE BLACK-SCHOLES EQUATION ‖

The state of the stock market displayed on a trader's screen is history. Big changes can occur in the fraction of a second that it takes for information to reach the screen. Computers have replaced people in executing deals. Humans are so slow that they are irrelevant for many market functions. In today's electronic trading, the markets are driven by computers rather than by people.

The software to operate in the new conditions is designed by quantitative analysts or 'quants'. They specialise in applying complex mathematical models to analyse market trends and execute trades. The computer-based algorithms, or 'algos', use all available relevant data – far beyond the analytical capacity of humans – to predict prices. Financial organisations now employ staff who have no background in finance but who have excellent problem-solving skills. Many have PhDs in mathematics, physics or engineering. The methods used include stochastic differential equations, numerical analysis and complexity theory.

A mathematical equation published in 1973 by Fischer Black and Myron Scholes led to an explosive growth in trading of financial products called derivatives. These are basically options to buy or sell on a fixed date at an agreed price. Black and Scholes assumed that changes in stock prices have two components: an upward or downward trend; and random fluctuations or jiggles. The jiggles determine market volatility and their size affects the value of the derivative.

The Black–Scholes equation is a partial differential equation, similar to the heat equation in physics. This is no surprise: the molecules of a gas move in random fashion, with innumerable jiggles – like market prices – so their mean behaviour, or temperature, is governed by a similar equation.

$$\frac{\partial V}{\partial t} + \frac{1}{2}\,\sigma^2 S^2 \frac{\partial^2 V}{\partial S^2} + rS\frac{\partial V}{\partial S} = rV$$

The Black–Scholes equation.

Black and Scholes assumed that volatility follows the so-called normal distribution or bell-shaped curve. The bell curve is found throughout science and economics. For example, if you toss a coin many times, heads and tails should show up roughly an equal number of times. The likelihood of a particular number of heads follows the bell curve. The crucial factor is independence: each coin toss is unaffected by the previous ones.

The Black–Scholes equation enabled analysts to reduce investment risk and increase profits using mathematics. But on several occasions following the equation's emergence, major market upheavals showed its limitations. The most dramatic of these was in 2007, when the housing bubble burst and the markets collapsed. Global shockwaves left the world economy reeling and the consequences are still being felt. Government intervention was need to bail out banks that were 'too big to fail'.

A common factor of the turbulent episodes was the inability of mathematical models to simulate highly volatile markets. The bell curve is fine for modelling numerous uncorrelated events. But if investors stop acting independently and follow their herd instincts, greed induces reckless behaviour or fear leads to panic. Extreme conditions ensue and they are beyond the range of validity of simple models like Black–Scholes.

Mathematicians are working to develop refined models and equations that give more realistic descriptions of market behaviour. They include intermittent surges that fatten the tails of the bell curve, making extreme conditions much more likely. With enormous profits to be made, these efforts will continue. In markets that are becoming ever more complex, mathematical models are essential, but they must be used responsibly and with due recognition of their limitations.

Given human frailty and irrationality, there will always be surprise crashes and booms. Even the great Isaac Newton was caught offside in 1720 by the infamous South Sea Bubble, one of the early market crashes. He lost more than £20,000, more than a million pounds today, and is said to have remarked, 'I can calculate the movements of the stars, but not the madness of men.'

ECCENTRIC PIZZA SLICES

Suppose six friends visit a pizzeria and have enough cash for just one big pizza. They need to divide it fairly into six equal pieces. That's simple: cut the pizza in the usual way into six equal sectors. But suppose there is meat in the centre of the pizza and three of the friends are vegetarians. How can we cut the pizza into slices of identical shape and size, some of them not including the central region?

More formally, here is the mathematical problem to be solved: *Divide a disc into a finite number of congruent pieces in such a way that not all of them touch the centre.* Have a think about this before reading on. There is more than one solution.

Clearly, since some of the six pieces must contain arcs of the outer edge, all pieces must have similar arcs on their boundaries. Drawing arcs of the same radius as the pizza, with centres on the edge, we get the pattern on the left below. Each of the six pieces is a curvy triangle, reminiscent of a yacht's spinnaker. All pieces are congruent – that is, identical in shape and size.

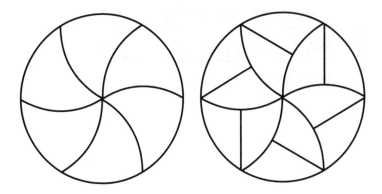

But this does not solve the problem, since all six slices touch the centre. But now we bisect each slice, as shown on the right. Then we get twelve pieces, all identical in shape and size, six of which exclude the central region. Now the three vegetarians can each choose two outer pieces, leaving the inner ones for the carnivores.

The solution we have shown is the basis for the logo of the Mathematics Advanced Study Semesters (MASS) program at Penn State University, which takes place annually every autumn (see www.math.psu.edu/mass/).

Problem: In addition to the solution shown above, there is an infinite family of other solutions. Can you find them? *Hint:* Start from the pattern in the left-hand panel above and sub-divide each piece.

MERCATOR'S MARVELLOUS MAP

Try to wrap a football in aluminium foil and you will discover that you have to crumple up the foil to make it fit snugly around the ball. In the same way, it is impossible to represent the curved surface of the earth on a flat plane without some distortion. To depict a region on a plane map, we have to project the spherical surface onto the flat plane.

Cartographers have devised many projections, each having advantages and drawbacks. No single projection can preserve all properties of the earth. Some keep areas the same but distort angles; others do the reverse. Some preserve distances, others preserve directions. Some are suitable for polar regions, others for the tropics. No one projection is ideal; there is no perfect map.

One very desirable property of a projection is that of preserving shapes, at least for small regions. If two curves on earth intersect at a certain angle, the corresponding 'image' curves on the map should intersect at the same angle. For example, parallels and meridians are perpendicular on the globe, so they should remain so on the map. Projections which are shape-preserving are called conformal. Many popular map projections are *conformal*.

For a sailor plotting a course, it is convenient to maintain a constant direction or compass bearing. Such a course is called a loxodrome or rhumb line. It is not the shortest route between two points (a great circle is), but it makes life easy for the navigator.

In 1569 the Flemish geographer Gerardus Mercator devised a map on which loxodromes appear as straight lines. This map quickly became the standard for nautical purposes. Navigators could now plot a loxodromic course as a straight line joining their departure point and destination, and follow this course across the ocean by maintaining a fixed bearing.

On Mercator's projection, all parallels of latitude have equal length. Since on earth their lengths decrease towards the poles, the parallels are stretched on the map. To ensure a conformal projection, the north–south spacing between them must also increase with latitude, expanding areas in polar regions. The projection is based on a cylinder tangent to the earth at the equator.

In the tropics, the cylinder is close to the earth's surface, and an accurate representation results. At high latitudes, the cylinder departs indefinitely far from the earth: distances and areas are grossly distorted. As a result, Mercator's map becomes useless at high latitudes due to the excessive distortion.

But take a look at an Ordnance Survey of Ireland map: it is a transverse Mercator projection, based on an Airy modified spheroid. For a regular Mercator map, the cylinder touches the earth along the equator. For a transverse projection, the cylinder is turned through a right angle, so it is tangent to the earth along a meridian of longitude. Therefore, distortion is minimised along this central longitude.

For Ireland, the central meridian is 8 degrees west, a line extending from Horn Head in County Donegal, southward through Lough Ree in the centre of Ireland and reaching Ballycotton in east County Cork. This is the only meridian that appears as a straight line on the OSI maps. These maps have minimal distortion in the vicinity of Ireland, the region of maximum interest. Mercator's ingenious invention – so beneficial to the early navigators and explorers – is still a powerful boon to us today.

THE REMARKABLE POWER OF SYMMETRY ‖

The number of women who have excelled in mathematics is lamentably small. Many reasons may be given, foremost being that the rules of society well into the twentieth century debarred women from any leading role in mathematics and indeed in science. But a handful of women broke through the gender barrier and made major contributions.

Perhaps the earliest was Hypatia, who taught philosophy and astronomy at the Neoplatonic school in Alexandria. In AD 415 she became embroiled in a feud and was murdered by an angry mob. Other noteworthy female mathematicians include Sophie Germain and Sofia Kovalevskaya. Probably the most brilliant of all was Emmy Noether, born in Erlangen, Germany in 1882.

Noether was the daughter of a professor of mathematics at the University of Erlangen. She must have learned maths with the help of her father, for she was excluded from access to any higher-level teaching. Through personal study and research, she became an expert in the theory of invariants, quantities that retain their value under various transformations.

The conservation of energy is a fundamental principle of science. Energy may take different forms and may be converted from one form to another, but the total amount of energy remains unchanged. Around 1915, when Albert Einstein was putting the final touches to his theory of general relativity, two mathematicians in Göttingen, David Hilbert and Felix Klein, became concerned about a problem in the theory: energy was not conserved. They felt that, given her knowledge, Noether might be able to solve the problem, so they invited her to come to Göttingen.

She accepted with enthusiasm: Göttingen was the leading mathematical centre and Hilbert the leading mathematician at that time. Hilbert made efforts to persuade the university authorities to hire Noether, but got her only an unpaid teaching post. However, she had greater success, coming up with a truly remarkable theorem that relates conserved quantities and symmetries.

It is usually surprising and occasionally delightful when apparently unrelated concepts or quantities are found to be intimately connected. Energy is generally conserved in physical systems. Under certain circumstances, so is angular momentum, roughly the *spin* of a body. And there are several other conserved quantities.

The mathematical expression that encapsulates the dynamics of a system is called the Lagrangian, after the outstanding French mathematician Joseph-Louis Lagrange. If a change of a basic variable, such as the position of the system or a shift of the time origin, leaves the Lagrangian unchanged, we have a symmetry. Noether found a totally unexpected connection between conserved quantities and symmetries of the Lagrangian.

Noether's Theorem does much more than simply establish a relationship between symmetries and conserved quantities. It provides an explicit formula by means of which, knowing a symmetry, we can actually calculate an expression for the quantity that is conserved. Moreover, the theorem was not confined to the classical mechanics of Newton, but found its true potential when used in the context of quantum mechanics.

With the rise of the National Socialist Party, Noether, along with many others, was dismissed from Göttingen in 1933. She emigrated to America, taking a position at Bryn Mawr College in Pennsylvania. Sadly, she died just two years later at the height of her creative powers. Her remarkable standing and reputation can be seen from the obituary written by Einstein in the *New York Times*: 'Fräulein Noether was the most significant creative mathematical genius thus far produced since the higher education of women began.'

INCREASINGLY ABSTRACT ALGEBRA

In the seventeenth century, the algebraic approach to geometry proved to be enormously fruitful. When René Descartes (1596–1650) developed co-ordinate geometry, the study of equations (algebra) and shapes (geometry) became inextricably interlinked. The move towards greater abstraction can make mathematics appear more abstruse and impenetrable, but it brings greater clarity and power, and can lead to surprising unifications.

From the mid-nineteenth century, the focus of mathematical study moved from numbers and the solution of individual equations to consideration of more general structures such as permutations and transformations. In studying whether equations higher than the fourth degree could be solved, Évariste Galois had shown the connection between groups of transformations and the roots of polynomials. Attention moved towards transformation groups and then towards more abstract groups, and a variety of other structures, like rings and fields.

The link between discrete (finite) groups and polynomial equations made by Galois inspired Sophus Lie to seek a similar link between infinite (continuous) groups and differential equations. The theory of Lie groups and Lie algebras that emerged from this has had a profound impact on mathematics. A Lie group is an algebraic group and also a topological manifold (a space that, locally, looks like Euclidean space). Moreover, the two aspects are entwined: the algebraic group operations are continuous in the topology.

ADVANCING ABSTRACTION

Through the twentieth century, the trend towards greater abstraction continued. One of the greatest contributors to this movement was the outstanding German mathematician Emmy Noether. Following Noether's abstraction of algebra, the nature of algebraic geometry was profoundly changed.

Emmy Noether has been called 'the mother of modern algebra'. The renowned algebraist Saunders Mac Lane wrote that abstract algebra started with Noether's 1921 paper 'Ideal theory in rings' and Hermann Weyl said that she 'changed the face of algebra by her work'. Noether's algebraic work was truly ground-breaking and hugely influential.

Continuous transformation groups are intimately related to symmetry, which is a powerful organising principle. Noether used the theory of Lie groups to derive her theorems relating symmetries of the Lagrangian to conserved quantities of a dynamical system (see page 275). Noether's theorems play an important role in dynamics and quantum mechanics.

Whole numbers can be added and multiplied. Systems with these two operations came to be known as rings. Many other examples of rings were found, comprising objects other than numbers. For example, the set of even numbers is a ring and the set of all polynomials in the variable x with integer coefficients is a ring.

In the mid-twentieth century, Oskar Zariski studied shapes called varieties, which are associated with rings. Every variety was associated with a ring, but the converse connection was not clear: not every ring corresponded to a variety. The extraordinary and eccentric genius Alexander Grothendieck found the connection by introducing new abstract structures called schemes, and revolutionised the subject of algebraic geometry. Varieties and schemes are central to the modern research in algebraic geometry. They are the abstract offspring of the geometric shapes and algebraic equations unified by Descartes so long ago.

ACOUSTIC EXCELLENCE AND RT-60 ‖

Attending a mathematical seminar in UCD recently, I could understand hardly a word. The problem lay not with the arcane mathematics but with the poor acoustics of the room. The lecturer spoke so rapidly that reverberation reduced his words to an indecipherable aural stew.

Sound indoors depends dramatically on the acoustic properties of the space. The sound is reflected by hard surfaces and absorbed by soft ones, gradually dying away as its energy is dissipated. Reverberation, or *reverb*, is particularly noticeable when a sudden bang continues to be audible for several seconds.

Shortly after the Fogg Lecture Hall at Harvard University was inaugurated in 1895, problems became clear: the acoustics were so poor that the audience couldn't understand what the lecturers were saying. Excessive reverb confused their voices. The problem was presented to a young physics professor, Wallace Clement Sabine.

Sabine approached the issue scientifically: he measured the sound under various circumstances using a pipe organ and a stopwatch. He noted the time lapse between stopping the source and the sound becoming inaudible. This corresponded to a reduction in amplitude by a factor of one thousand, or in sound intensity by a factor of a million. In quantitative terms, it was a power drop of 60 decibels.

Sabine developed a formula for the reverberation time. He found that it is proportional to the size of the room and inversely proportional to the total absorption. Sabine's reverberation equation is

$$\mathrm{RT}_{60} = 0.16 \, V / a \, S$$

where V is the volume of the room, S is the total surface area and a the absorption coefficient. When all other quantities are in standard metric units, the reverberation time RT_{60} is given in seconds.

A hard surface, like a painted wall, has little absorption, so a is small, while an open window produces no reflection at all and so is a perfect absorber ($a = 1$). The total absorption changes with frequency. The shape of the room and the damping effect of the air can also be important.

For Fogg Hall, Sabine measured a reverberation time of more than five seconds. For a lecture hall, a short reverberation time is required. If a syllable continues to sound after the next one is spoken, confusion will result. The materials used for the walls, ceiling, floor and furnishing all affect reverb. The presence of people in the auditorium also changes the acoustics. After many experiments, Sabine found a solution: using sound-absorbing materials on the walls, he converted the hall to a space suitable for lecturing.

Sabine is considered the father of architectural acoustics. Architects must consider reverberation when designing spaces, so that enclosures have acoustic properties suitable for their purpose. Acoustic architects can produce reference sounds, record them on digital sound metres and use Fourier transform methods to measure the acoustic response of an auditorium. Thanks to scientific developments, it is now possible to design a space that is suitable for lectures, quartets, opera and a full orchestra.

Sabine's success with Fogg Hall was the making of his career. He went on to act as consultant for the Boston Symphony Hall, the first concert hall to be designed using quantitative acoustics. It opened in 1900 and is still regarded as one of the best concert halls in the world.

THE BRIDGES OF PARIS

Leonhard Euler, the outstanding Swiss mathematician, considered a problem known as the Seven Bridges of Königsberg. It involves a walk around the city now known as Kaliningrad, in the Russian exclave between Poland and Lithuania. Since Kaliningrad is out of the way for most of us, let's go closer to home and have a look at the bridges of Paris.

In the centre of Paris, two small islands, Île de la Cité and Île Saint-Louis, are linked to the Left and Right Banks of the Seine and to each other. We consider only bridges that link the two islands to each other or to the river banks, and ignore all other bridges spanning the Seine. The figure shows that there are seven bridges from the Left Bank, seven from the Right Bank, ten from Île de la Cité and six from Île Saint-Louis. This makes thirty in all, but as each bridge is double-counted, there are in total fifteen bridges.

Inspired by Euler, let us seek walking routes through the centre of Paris that cross each bridge exactly once. We distinguish between *open* routes that link from one point to another, and *closed* routes that lead back to the starting point. We pose three problems. The first is this: starting from Saint-Michel on the Left Bank, walk continuously so as to cross each bridge exactly once. The second problem is not so simple: start at Notre-Dame Cathedral, on Île de la Cité, and cross each bridge exactly once. The third problem is this: find a closed route crossing each bridge once and returning to its starting point.

Euler reduced the problem to its basics, replacing each piece of land by a point or *node* and each bridge by a *link* joining two nodes. He saw immediately that any node not at the start or end of the walk must have an even number of links: every link into the node must have a corresponding exit link. Thus, there can be at most two nodes with an odd number of links. Moreover, if any node has an odd number of links, there are no closed routes.

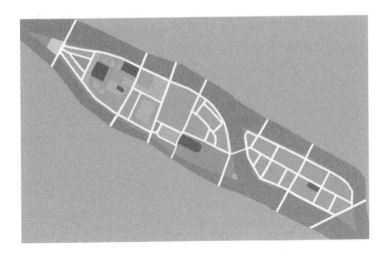

For the Paris problems, both banks of the Seine have seven bridges, an odd number. The two islands each have an even number of links. Thus, the Left and Right Banks must be at the start and end of the route. It is simple to find a route starting at Saint-Michel and ending at Hôtel de Ville on the Right Bank. So, the first problem is easily solved.

Since there are an even number of bridges to Île de la Cité, Notre-Dame Cathedral cannot be at the start or end of the route: the second problem has no solution. Finally, since both banks must be termini, there is no closed route and the third problem is also impossible to solve.

Euler's original problem, the Seven Bridges of Königsberg, had no solutions at all, as each of the four nodes had an odd number of links. His problem, discussed in virtually every textbook on graph theory, is considered the first real problem in topology. This branch of mathematics is concerned with continuity and connectivity, and studies properties that remain unchanged under continuous deformations, such as stretching or bending but not cutting or gluing. Topology is a powerful unifying theme in modern mathematics.

BUFFON WAS NO BUFFOON ||

One of the earliest problems in geometric probability was posed and solved by Georges-Louis Leclerc, Comte de Buffon in 1733. He considered the question of dropping a stick of length L onto a wooden floor with floorboards of width D. Two outcomes are possible: the stick may land on a single board, or it may cross the gap between two boards.

Buffon regarded this process as a game of chance and he was interested in the odds of a given drop crossing a line. Since the game can be played on a table top with a needle and a ruled sheet of paper, it is generally known as Buffon's Needle.

For simplicity, we may assume that $L = D = 1$; this assumption is not limiting. If the stick is dropped 'at random', it is fairly easy to show that the probability that it crosses a line between boards is $p = 2/\pi$.

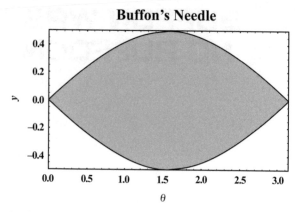

Buffon's Needle

The shaded region corresponds to crossings of the needle.

ESTIMATING π

Now if we drop a stick a large number of times n, and get k crossings, we can estimate the probability by means of the frequency ratio k/n, so

$$(2/\pi) \approx (k/n) \quad \text{or} \quad \pi \approx 2n/k$$

But there is a big problem: for a large number of trials, the standard error ε is inversely proportional to the square root of n, so the number of trials increases as $1/\varepsilon^2$.

If we want a result accurate to three significant figures, we might let $\varepsilon = 10^{-3}$, but this implies $n \approx 10^6$. We need something like one million throws to get π to three significant digits – ignoring factors of order one. For six-digit accuracy, we need a trillion throws.

The table shows estimates and percentage errors computed for n in {10, 100, 1000, 10000, 100000}. It is clear that the accuracy is increasing with n, but painfully slowly. One million terms yield only three significant figures accuracy.

n	Estimate of π	Error (%)
10	2.5	−20.4225
100	2.8169	−10.3353
1,000	3.19489	1.6965
10,000	3.11721	−0.7761
100,000	3.13239	−0.2929
1,000,000	3.14433	−0.0871

Why did Buffon propose such an impractical method of estimating π? He did not!

In a 2014 article, Ehrhard Behrends found no evidence that Buffon did the 'experiment'. He was interested in the odds of a game of chance, assuming π to be known. Laplace realised that Buffon's experiment could, in principle, be used to estimate π, but there is no indication that he tried this. Many experiments with Buffon's Needle are documented. In one, Lazzerini claimed to have obtained the value 355/113 for π. This is accurate to seven figures but, as only 3408 throws were involved, it cannot be given credence.

Many variations on Buffon's Needle have been proposed. One especially curious result is called Buffon's Noodle. If the needle is bend into any (plane, rectifiable) curve, the probability of crossing a line remains unchanged. This perplexing result is less incredible when it is realised that multiple crossings for a single throw are now possible, and must be counted accordingly. But it remains a surprising result.

JAMES JOSEPH SYLVESTER

James Joseph Sylvester was born in London to Jewish parents in 1814. The family name was Joseph but, for reasons that are unclear, Sylvester – the name of an anti-Semitic Pope from the Roman period – was adopted later.

Sylvester's mathematical talents became evident at an early age. He entered Cambridge University in 1831, aged just seventeen, and came second in the notorious examinations known as the Mathematical Tripos; the man who beat him achieved nothing further in mathematics! Trinity College Dublin awarded Sylvester the degrees of BA and MA that Cambridge would not bestow unless he subscribed to the Thirty-Nine Articles of the Church of England, which his Jewish convictions precluded. He was elected a Fellow of the Royal Society at the unusually early age of 25.

Sylvester, like many brilliant young people today, had difficulty in finding a suitable position. He took a post in Virginia, but it turned out badly and he returned from

America within a year or so. He supported himself by taking private pupils, the most distinguished of these being Florence Nightingale.

In 1846, aged 32, Sylvester entered the Inner Temple to pursue a legal career, and was called to the Bar four years later. The greatest benefit of this period was his encounter with Arthur Cayley, another renowned mathematician, which led to a lifelong friendship and collaboration. Cayley and Sylvester inspired each other in some of their best work, on the theory of invariants and matrix theory.

The theory of algebraic invariants had initially been developed by George Boole while he was at Queen's College, Cork. At school we learn to solve quadratic equations like $ax^2 + bx + c = 0$. Quadratics have two solutions, called roots. The nature of the solutions depends on a quantity $b^2 - 4ac$, which is called the discriminant (Sylvester introduced this term). If the discriminant is equal to zero, the two roots are equal. This property remains unchanged under a range of modifications called bilinear transformations: the discriminant is an *invariant*. Boole had found that such invariants exist for all algebraic equations. Invariance under transformations is fundamental in modern mathematics and also plays a key role in the physical sciences.

In 1854 Sylvester applied for the professorship of mathematics at the Royal Military Academy, Woolwich. He was unsuccessful and the position went to Ennis-born Matthew O'Brien, one of the founders of vector analysis. But following O'Brien's death the following year, Sylvester was appointed. After retiring from Woolwich, he took up a professorship in Johns Hopkins University in Baltimore, established in 1876. There he founded the *American Journal of Mathematics*, among the most celebrated and longest-running mathematical journals. The journal gave American mathematics a tremendous impetus, and has been in continuous publication ever since.

In 1883 the brilliant Irish number theorist Henry John Stephen Smith, who had occupied the Savilian Chair of Geometry at Oxford since 1861, died aged only 57. Sylvester, aged 70, accepted the offer of the vacant chair and held it until his death in 1897.

Sylvester had a broad and deep knowledge of classical literature, and his mathematical papers are peppered with Latin and Greek quotations. He also wrote poetry and a book, *The Laws of Verse*. He had a great interest in music, once taking singing lessons from Gounod. Sylvester wrote: 'May not music be described as the mathematics of the sense, mathematics as music of the reason? The musician feels mathematics, the mathematician thinks music: music the dream, mathematics the working life.'

HOLBEIN'S ANAMORPHIC SKULL ‖

Hans Holbein the Younger, court painter during the reign of Henry VIII, produced some spectacular works. Among the most celebrated is a double portrait of Jean de Dinteville, French Ambassador to Henry's court and Georges de Selve, Bishop of Lavaur. Painted by Holbein in 1533, the picture, known as *The Ambassadors*, hangs in the National Gallery, London (www.nationalgallery. org.uk/paintings/hans-holbein-the-younger-the-ambassadors).

Diagonally across the lower foreground of *The Ambassadors* there is an elongated form, the significance of which is not immediately obvious. It is a *perspective anamorphosis*, an intentionally distorted image whose nature becomes clear when it is viewed from an oblique angle. From a vantage point to the centre right of the picture, the form reveals itself to be a skull, probably included as a *memento mori*.

PERSPECTIVE PAINTING

Visual artists developed the techniques of painting in perspective during the early Renaissance, and mathematicians developed projective geometry, which systematises the principles of perspective. Holbein had a deep appreciation of these principles, but the precise method that he used to produce the anamorphic form is still a matter of debate.

The picture contains images of books, globes and scientific apparatus, reflecting the learning of the two wealthy and influential subjects. It is approximately life size, painted in oil on an oak panel, and is close to an exact square in shape. Detailed analysis of the anamorphic form has been done, using a range of complex mathematical transformations.

Computer graphics can be used to reveal the intention of the artist. A sequence of three transformations, a rotation R, a stretching transformation S and an inverse transformation R^T

$$T = R^T S R$$

produces an undistorted image of a skull. The three operations are equivalent to a single stretching transformation about an oblique axis. A precise analysis leads to the following image.

A high-resolution depiction of the anamorphic skull in Hans Holbein's The Ambassadors.

The uses of the technique of anamorphosis are not limited to the fine arts. As a modern application, advertisers display images flat on a football pitch which, when viewed obliquely through a television camera, appear to be vertically standing signs.

THE UBIQUITOUS CYCLOID

Imagine a small light fixed to the rim of a bicycle wheel. As the bike moves, the light rises and falls in a series of arches. A long-exposure nocturnal photograph would show a *cycloid*, the curve traced out by a point on a circle as it rolls along a straight line. A light at the wheel hub traces out a straight line. If the light is at the mid-point of a spoke, the curve it follows is a curtate cycloid. A point outside the rim traces out a prolate cycloid, with a backward loop.

Cycloids have been studied by many leading mathematicians over the past five hundred years. The name *cycloid* originates with Galileo, who studied the curve in detail. The story of Galileo dropping objects from the Leaning Tower of Pisa is well known. Although he could not have known it, a falling object traces out an arc of an inverted cycloid. This is due to the tiny deflection caused by the earth's rotation. Moreover, an object thrown straight upwards follows the loop of a prolate cycloid, landing slightly to the west of its launch point.

Blaise Pascal, who had abandoned mathematics for theology, found relief from a toothache by contemplating the properties of cycloids. Taking this to be a sign from above, he resumed his mathematical researches. Pascal proposed some problems on the cycloid and one of the respondents was Christopher Wren, best known as the architect of St Paul's Cathedral in London. Wren proved that the length of a cycloid arch is four times the diameter of the circle that generates it. Today, this is an easy problem in integral calculus but in 1658 it was a formidable achievement.

In 1696, Johann Bernoulli posed a problem that he called the *brachistochrone* – or shortest time – problem: find the path along which gravity brings a mass most quickly from one point to another one not directly below it. The five mathematicians who responded included Newton, Leibniz and Johann's brother Jakob. The desired path is a cycloid.

The story goes that Newton received the problem one evening upon returning from the Royal Mint, where he was Master. He stayed up late working on it and by 4 a.m. he had obtained a solution, which he mailed later that morning. Although his solution was anonymous, Bernoulli immediately perceived its authority and brilliance, giving his reaction in the classic phrase '*ex ungue leonem*', 'the lion is recognised by his claw'.

Cycloid arches have been used in some modern buildings, a notable example being the Kimbell Art Museum in Fort Worth, Texas, designed by the renowned architect Louis Kahn. Like many classical buildings, the museum is based on a consistent mathematical model. The basic plan is composed of cycloid vaults arranged in parallel units. These vaults have gently rising sides, giving the impression of monumentality. This geometric form is capable of supporting its own weight and can withstand heavy pressure.

In the atmosphere, the rotation of the earth generates cycloidal motion: icebergs and floating buoys have been seen to trace multiple loops of a prolate cycloid. Finally, epicycloids and hypocycloids are used in modern gear systems as they provide good contact between meshed gear teeth, giving efficient energy transmission.

Cycloids: Common (black), Curtate (dark grey) and Prolate (light grey)

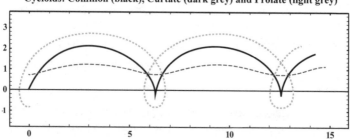

HAMMING'S SMART ERROR-CORRECTING CODES

In the late 1940s, Richard Hamming, working at Bell Labs, was exasperated with the high level of errors occurring in the electro-mechanical computing equipment he was using. Punched card machines were constantly misreading, forcing him to restart his programs. He decided to do something about it. This was when error-correcting codes were invented.

A simple way to detect errors is to send a message twice. If both versions agree, they are probably correct; if not, there is an error somewhere. But the discrepancy gives us no clue as to where the error lies. Sending the message three times is better: if two versions agree, we assume they are correct and ignore the third version. But there is a serious overhead: the total data transmitted are three times the original volume; the information factor is 1/3.

THE (7,4) CODE

Hamming, an American mathematician and computer scientist, devised a much more efficient set of codes, now called Hamming Codes. The simplest of these is the (7,4) code. The message to be sent is split into groups of four binary digits or bits, denoted d_1, d_2, d_3 and d_4. Then three additional check-bits or *parity bits* p_1, p_2 and p_3 are added, so the message becomes $d_1d_2d_3d_4p_1p_2p_3$ with information content 4/7.

These parity bits are defined in such a way that, if any single bit is in error, it can be identified and corrected. In the Venn diagram opposite, the data bits and parity bits are shown in different regions of the diagram. The parity bits are chosen so that there is an even number of ones in each circular region. This can always be done.

If any single bit is in error (a zero instead of a one, or vice versa), the evenness or parity is disrupted. Each of the seven bits destroys parity in a different set of circles, so the faulty bit can be detected and corrected. This is the genius of Hamming's code: we know which bit is in error.

Of course, if the message is badly contaminated, there may be more than one error in the string of seven bits. We may be able to detect this, but not to correct it. More refined encodings are required for this. One such, deriving from Hamming's code, is SECDED (single error correction, double error detection), widely used in computer memories.

Coding theory is an active field of mathematical research today. We depend on reliable communication channels that transmit large volumes of data. These data must be compressed before sending and accurately expanded on arrival. If it is sensitive, it must be encrypted. And inevitable errors in noisy transmission channels must be detected and corrected.

Many clever codes have been devised, such as the Reed–Solomon code, used to eliminate noise in CD recordings. Richard Hamming's wrestling match with punched card equipment has led to a worldwide industry.

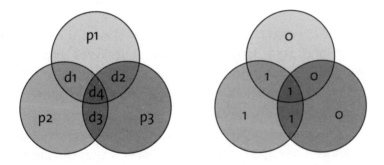

Venn diagram with the four data bits d_1 to d_4 and three parity bits p_1 to p_3.

MOWING THE LAWN IN SPIRALS

Broadly speaking, a spiral curve originates at a central point and gets further away as it revolves around the point. Spirals abound in nature, being found at all scales from the whorls at our fingertips to vast rotating spiral galaxies. The seeds in a sunflower are arranged in spiral segments. In the technical world, the grooves of a gramophone record and the coils of a watch balance-spring are spiral in form.

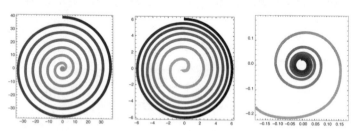

Left: Archimedean spiral. Centre: Fermat spiral. Right: hyperbolic spiral.

In polar co-ordinates, the radial distance $r(\vartheta)$ from the central point is a monotonic function of the azimuthal angle ϑ. There are several canonical spiral forms. The simplest is the Archimedean spiral, $r = a\,\vartheta$. This is generated by a point moving with uniform speed along a ray that is rotating with constant angular speed. Since the points of intersection with a fixed ray from the origin are evenly spaced with separation $2\pi a$, it is also called an arithmetic spiral (left-hand panel of the diagram). It was described by Archimedes in his work *On Spirals*.

More generally, we consider $r = a\vartheta^k$. For $k = \frac{1}{2}$ we get Fermat's spiral $r = a\sqrt{\vartheta}$ (centre panel of the diagram). For $k = -1$ we have $r = a/\vartheta$, which is a hyperbolic spiral (right-hand panel of the diagram).

An equiangular spiral is such that every ray from the origin cuts it at the same angle. Its equation is $r = a$ exp $(b\,\vartheta)$. Since $b\,\vartheta = \log(r/a)$, it is also called a logarithmic spiral and, since the intersection points with a fixed ray form a geometric sequence, the name geometric spiral is also used. Christopher Wren observed that many sea shells, such at the nautilus, have logarithmic spiral cross-sections.

EASY GRASS-CUTTING

Suppose you wish to cut the grass. Here is an easy way:

- Erect a stout column in the centre of the lawn.

- Tie the mower to the column with a long rope.

- Start it so that it winds inwards in a spiral arc.

- Relax and enjoy the magic of the automower.

The curve traced by the mower looks like an Archimedean spiral. It is actually slightly different: it is the involute of a circle (namely, the circular cross-section of the central column). The column should be chosen such that $2 \pi a = D$ where D is the blade diameter of the mower.

If the column is described by the equations

$$x = a \cos \vartheta \qquad y = a \sin \vartheta$$

where a is the column radius, then the curve traced out by the mower is

$$x = a \left(\cos \vartheta + \vartheta \sin \vartheta \right) \qquad y = a \left(\sin \vartheta - \vartheta \cos \vartheta \right)$$

The radius vector is $r^2 = a^2 \left(1 + \vartheta^2 \right)$. For $\vartheta = 0$ we have $r = a$, whereas for the Archimedean spiral ($r^2 = a^2 \vartheta^2$) we have $r = 0$ when $\vartheta = 0$. The two curves are close, but not identical.

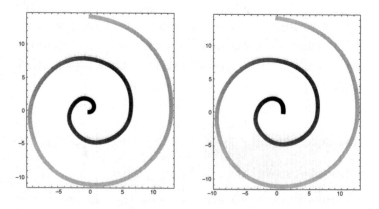

Left: Archimedean spiral. Right: involute of a circle.

MELENCOLIA I: AN ENIGMA FOR HALF A MILLENNIUM ‖

Albrecht Dürer, master painter and engraver of the German Renaissance, made his *Melencolia I* in 1514, just over five centuries ago. It is one of the most brilliant engravings of all time, and among the most intensely debated works of art. The winged figure, Melancholy, sits in a mood of lassitude and brooding dejection, weighed down by intellectual cares. Her head rests on her left hand while her right hand holds a mathematical compass, one of many symbols and motifs in the work that reflect Dürer's interest in mathematics.

This allegorical work is full of enigmas. It may depict Melancholy as the inevitable counterpart of creativity. Some art experts have described it as a spiritual self-portrait of Dürer. The artists of the early Renaissance strove to understand the natural world that they were depicting. Given the primitive state of science at the time, they were driven to develop knowledge of anatomy, geometry and optics. Indeed, some profound scientific advances were initiated by the investigations of visual artists.

Dürer was conversant with the principles of perspective; this is clear from *St Jerome in his Study*, an engraving made in the same year as *Melencolia*. He wrote *The Painter's Manual* to guide young artists in the geometric principles underlying representational art.

Behind the main figure is a four-by-four grid with the first sixteen counting numbers. Each row and each column sums to 34, as do the diagonals. This is the first occurrence in Western art of a magic square. It has many other interesting mathematical symmetries. A curious aspect of the square is the appearance of the numbers 15 and 14 in the centre cells of the bottom row, corresponding to 1514, the year the engraving was made.

On the left side of *Melencolia* we see a strange-looking rainbow above a marine horizon and, off-centre behind it, a comet with blazing rays. This may have been inspired by the comet observed at Christmas 1471, the year of Dürer's birth. The rainbow cuts the horizon vertically, implying that the source of light is near the horizon. It is narrower than a real rainbow, and has no striped pattern: it may be a moon-bow, since the comet suggests a nocturnal scene.

The large polyhedron at centre left has four visible faces, three pentagons and an equilateral triangle. Assuming that the hidden side of the polyhedron is of similar form, we have six pentagons and two triangles. It is a truncated rhombohedron, like a stretched cube with two corners clipped off, but just what inspired Dürer to draw it is a mystery. On the front of the polyhedron, the ghostly image of a face can be seen.

At bottom left is a spherical ball. Its illumination indicates the source of light, behind and to the right of the viewer. Some carpenter's or wood-turner's tools lie on the floor. On the wall behind the figures hang a balance, an hourglass, a sundial and a bell.

Examination of the high-resolution image of the engraving available on Wikipedia is a rewarding experience and will reveal a wealth of interesting details. There are many other enigmas in *Melencolia* and you may discover something new or dream up a novel theory about meanings hidden in the work. Perhaps you can explain the shadowy figure on the face of the polyhedron.

MATHEMATICS CAN SOLVE CRIMES

What use is maths? Why should we learn it? A forensic scientist could answer that almost all the mathematics we learn at school is used to solve crimes. Forensic science considers physical evidence relating to criminal activity and practitioners need competence in mathematics as well as in the physical, chemical and biological sciences.

Trigonometry, the measurement of triangles, is used in the analysis of blood spatter. The shape indicates the direction from which the blood has come. The most probable scenario resulting in blood spatter on walls and floors can be reconstructed using trigonometric analysis. Such analysis can also determine whether the blood originated from a single source or from multiple sources.

Suppose a body is found at the foot of a block of flats. Was it an accident, suicide or murder? Using the measured distance from the building, together with some elementary geometry and dynamics, the forensic scientist can form an opinion as to whether the victim fell, jumped or was pushed.

Ballistics calculations, like computing the ricochet angle of a bullet bouncing off a solid surface, use trigonometry. Bullet trajectories determine the distance from shooter to target and perhaps the height of the shooter and where they were standing when they shot the victim.

The exponential and logarithmic functions, found throughout science, play a key role in forensics. The exponential function relates to processes that depend on the amount of material present as time changes. Rates of heating or cooling, or of the metabolism and elimination of alcohol and drugs, are governed by exponential rates of change.

After death, a body cools until it reaches the environmental temperature. By modelling the heat loss mathematically, using Newton's law of cooling, an exponential decay of temperature difference is found. This enables estimation of the time elapsed since death. In practice, more elaborate models can be used.

For quantities that vary over many orders of magnitude, like the concentration of chemicals in the body, the logarithmic function allows us to compress them into a more manageable range. The pH scale, which indicates the level of acidity, is of this sort and is often vital in forensic work.

Probability and statistics are of growing importance in law enforcement throughout the world. Quantitative statistical analysis is used to compare sets of experimental measurements, to determine whether they are similar or distinct. This applies to glass fragments, drug samples, hairs and fibres, pollen grains and DNA sequences.

The chance of two people having identical DNA profiles may be one in a hundred trillion, but forensic biologists must work with crime scene samples that are small and degraded. This makes the identification of a unique individual more difficult, and subtle probabilistic arguments are required to draw inferences.

When analysing evidence from fingerprints, blood groups and DNA profiles, probability enters the scene. Conditional probabilities are of vital importance in a forensic context. Such considerations may determine whether two events relating to a crime are linked or independent. The calculation of the probability of occurrence of multiple events is subtle if the events are related. Failure to understand conditional probability has led to some tragic miscarriages of justice.

In conclusion, in the hands of the forensic scientist every piece of mathematics that we learn in school may prove to be a matter of life and death.

LIFE'S A DRAG CRISIS ‖

The character of fluid flow depends on a dimensionless quantity, the Reynolds number. Named for Belfast-born scientist Osborne Reynolds, it determines whether the flow is laminar (smooth) or turbulent (rough). Normally the drag force increases with speed.

The Reynolds number is defined as Re = VL/ν where V is the flow speed, L the length scale and ν the viscosity coefficient. The transition from laminar to turbulent flow occurs at a critical value of Re which depends on details of the system, such as surface roughness.

Normally the drag force increases with speed. In 1912, Gustave Eiffel – of Eiffel Tower fame – made a remarkable discovery, known as the *drag crisis*. Studying flow around a smooth sphere, he found a drop in the drag force as the flow speed increased above a critical Reynolds number 200,000 and continued to drop until about Re = 300,000 (see figure overleaf).

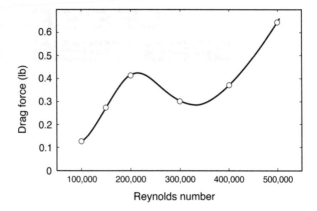

This is extraordinary: if we consider a spherical ball fixed in a wind tunnel, there is a point at which the drag force on the sphere actually decreases as the flow speed is increased.

The drag crisis was explained by Ludwig Prandtl in terms of his boundary layer theory. Reynolds had found that, as the speed increases, the flow changes from laminar to turbulent. The transition point occurs at the critical Reynolds number. Prandtl argued that the turbulence mixes rapidly moving external air into the boundary layer, increasing the range over which it adheres to the surface, making the trailing low pressure wake smaller and thereby reducing the drag force.

The drag crisis extends the range of a rapidly travelling ball such as a golf ball. For a smooth ball, the critical speed is well above the practically attainable range. By roughening the surface of the ball, the critical Reynolds number is reduced to about 50,000. The resulting reduction in drag can double the distance flown, compared to what is possible with a smooth ball.

THE FLIGHT OF ‖
A GOLF BALL

Golf balls fly further today, thanks to new materials and mathematical design. They are a triumph of chemical engineering and aerodynamics. They are also big business, and close to a billion golf balls are sold every year.

The golfer controls the direction and spin of the ball by variations in his stroke. A pro can swing his or her driver at up to 200 km/h, driving the ball 50% faster than this as it leaves the tee, on a trajectory about 10 degrees above the horizon. By elementary mechanics the vertical motion is decelerated by gravity, and the ball should bounce after about 200 metres and stop a few tens of metres further on.

How come, then, that professional golfers regularly reach over 300 metres? Gravity is not the only force acting on the ball, and aerodynamic forces can extend the range. In addition to drag, which slows down the flight, there is a lifting force that extends it. Drag is the force you feel when walking against a strong wind. Lift, the force that enables aeroplanes to fly, results from the back-spin of the ball. Air passing over the top of the ball is speeded up while air below is slowed. This results – by way of Bernoulli's principle – in lower pressure above and an upward force on the ball. Lift allows golfers to achieve greater distances.

As described in the preceding article, the Reynolds number determines whether the flow is laminar (smooth) or turbulent (rough). Normally the drag force increases with speed. However, as shown by Gustave Eiffel, above a critical Reynolds number, the drag force decreases with flow speed.

The drag crisis extends the range of a rapidly travelling ball, but a smooth ball needs speeds in excess of 300 km/h, unattainable in golf. So how can the drag crisis help? The answer lies in the dimples. The aerodynamics of a ball are determined by its mass and shape, in particular the nature of the surface. A complex pattern of dimples of various sizes, shapes and depths influences the air flow around the ball and causes a transition to turbulence at a lower Reynolds number. By roughening the surface, the critical Reynolds number is reduced to speeds within the golfer's range. The resulting decrease in drag can double the distance flown compared to what is possible with a smooth ball.

Dimples give golf balls greatly enhanced performance. Most balls have about 300 dimples. Manufacturers promise greater control, stability and velocity on longer shots. Hundreds of dimple patterns have been devised and patented, but the optimal design remains a challenge. How many dimples should there be and what shape and pattern should they have?

Simulating the flow around a golf ball by solving the governing equations (the Navier–Stokes equations) by numerical means is possible, but it requires hundreds of hours on a powerful computer cluster. Understanding the flight of a golf ball is still a challenge for applied mathematics, with no complete solution in sight.

FACTORIAL 52: A STIRLING PROBLEM

How many ways can a deck of cards be arranged? It is very easy to calculate the answer, but very difficult to grasp its significance.

There are 52 cards. Thus, the first one may be chosen in 52 ways. The next one can be any of the remaining 51 cards. For the third, there are 50 choices, and so on until just one card remains, leaving only the option to put it last.

Therefore, the total number of possibilities is

$$52! = 52 \times 51 \times 50 \times \ldots \times 3 \times 2 \times 1$$

This number is called factorial 52. To say that it is a large number is an understatement. The program Mathematica can compute to arbitrary precision and entering the command *Factorial*[52] yields the following result:

80658175170943878571660636856403766975289505440883277824000000000000

In more compressed notation, this is 8.06582×10^{67} or, to just a single figure of accuracy, 10^{68}; that is, 1 followed by 68 zeros.

DESCRIBING 52!

It is difficult to illustrate the size of 52! in terms of anything practical. People have talked about the number of drops in the ocean or how many grains of sand would fill the Grand Canyon. These numbers come nowhere close to 52!.

The number of atoms in the observable universe is estimated to be about 10^{80}, which is a trillion times bigger than 52!. But does this really help us to visualise what either of these numbers is like? The Wikipedia article on 'Names of Large Numbers' describes 10^{66} as an *unvigintillion*. Thus, $52! \approx 8 \times 10^{67}$ is about eighty unvigintillion. But this is just a name.

The universe is 4×10^{17} seconds old. If a random arrangement of cards were chosen each second during the entire life of the universe, only a tiny fraction of all possible orderings would be selected. The chance of the same ordering being chosen twice is utterly negligible. Even if a billion arrangements were chosen every second, there would still be no real chance of a duplicate.

STIRLING'S APPROXIMATION

The calculation of the number 52! is simple. Just multiply 52 by 51, the result by 50 and so on until you reach 1. But how tedious this is, and how error-prone! There is a beautiful expression giving an approximation to any factorial, named for James Stirling (1692–1770), a Scottish mathematician (although it seems that the result was stated earlier by Abraham de Moivre). The approximation is

$$n! \approx S_1(n) \equiv \sqrt{2\pi n}\left(\frac{n}{e}\right)^n$$

Plugging in the argument n = 52, the formula gives 8.0529 x 10^{67}, which is correct to two decimal places.

Another approximation was found among the papers of the Indian mathematician Srinivasa Ramanujan and published in his *Lost Notebook* in 1988:

$$\ln(n!) \approx n\ln(n) - n + \frac{1}{6}\ln(n(1+4n(1+2n))) + \frac{1}{2}\ln(\pi)$$

This gives 52! to one part in a billion.

SHUFFLING AND REPEATED ORDERS

With such a vast number of possibilities, one might ask if any randomly chosen order of a deck of cards occurs more than once. Making very reasonable assumptions, it is easy to argue that a particular ordering will never occur twice during the life of the universe. Thus, when you thoroughly mix up the cards, you are bound to arrive at an ordering that has never been seen before and will never be seen again.

However, there is a big proviso here. The shuffling of the cards must be sufficiently thorough to ensure true randomisation. Mathematical studies have indicated that a small number of effective shuffles suffice to mix up the pack to random order. Bayer and Diaconis showed that after seven random riffle shuffles, any of the 52! possible configurations is equally probable.

RICHARDSON'S FANTASTIC FORECAST FACTORY

Modern weather forecasts are made by calculating solutions of mathematical equations. These equations express the fundamental physical principles governing the atmosphere. The solutions are generated by complex simulation models with millions of lines of code, implemented on powerful computer equipment. The meteorologist uses the computer predictions to produce localised forecasts and guidance for specialised applications.

During World War I, long before the invention of computers, the English Quaker mathematician Lewis Fry Richardson devised a method of solving the equations and made a test forecast 'by hand'. The forecast was a complete failure, giving an utterly unrealistic prediction of pressure change, but Richardson's methodology was sound and it underlies modern computer weather forecasting.

In 1922 Richardson published a remarkable book, *Weather Prediction by Numerical Process*. Having described his method, he presents a fantasy of a 'Forecast Factory', a building with an enormous central chamber with walls painted to form a map of the globe. There a large number of (human) computers are busy calculating the future weather.

Richardson estimated that 64,000 people would be needed to calculate weather changes as fast as they were happening. In fact, this was over-optimistic: to produce a forecast in timely fashion would require upwards of a million computers.

The working of the Forecast Factory is co-ordinated by a Director of Operations. Standing on a central dais, he synchronises the computations by signalling with a spotlight to those who are racing ahead or lagging behind. There are striking similarities between Richardson's Forecast Factory and a modern massive parallel processor.

Several artists have created illustrations of the forecast factory. One particular image has recently come to light. The painting, in ink and watercolours, was made by Stephen Conlin in 1986, on the commission of Professor John Byrne, then Head of the Department of Computer Science in Trinity College Dublin. This painting, which has gone unnoticed for many years, is a remarkable work, rich in detail and replete with hidden gems.

Conlin's image depicts a huge building with a vast central chamber, spherical in form. On the wall of this chamber is a map with roughly half the globe visible. On an upper level sit four senior clerks. A banner on each desk identifies a major historical figure. Several scholars and savants are depicted in the painting: pioneers of computing including Charles Babbage, Ada Lovelace and George Boole and mathematicians like John Napier, Blaise Pascal and Gottfried Wilhelm von Leibniz.

Communication within the factory is via pneumatic carriers, systems that propel cylindrical containers through a pipe network using compressed air. These systems were used in large retail stores, such as Clerys in Dublin, to transport documents and cash. Pneumatic networks were popular in the late nineteenth and early twentieth centuries for transporting mail, documents or money within a building, or even across a city.

A description of Conlin's image can be found on the website of the European Meteorological Society: www.emetsoc.org/resources/rff/. There is a very high resolution version of the picture there, with a zoom facility. Examination of the high-res image is rewarding, and will reveal a wealth of interesting details.

THE ANALEMMATIC SUNDIAL

If you are ever in Dun Laoghaire, take a stroll out the East Pier and you will find an analemmatic sundial. In most sundials, the gnomon, the part that casts the shadow, is fixed and the hour lines radiate outwards from it to a circle. In an analemmatic sundial the hour points are on an ellipse, or flattened circle, the horizontal projection of a circle parallel to the equator. You yourself form the gnomon, and the point where you stand depends on the time of year. This is shown on a date scale set into the dial. Your shadow, falling somewhere on the ellipse, indicates the hour.

Advances in mathematics and astronomy have gone hand in hand for millennia. As civilisation developed, accurate time measurement became essential. More precise observations of the stars and planets called for more exact mathematical descriptions of the universe. The similarity between how we divide up angles and hours of the day arises from the use of astronomical phenomena to measure time. The division of a circle into

360 degrees, with each degree divided into 60 minutes and each minute into 60 seconds of arc, dates back to the Babylonians.

Tycho Brahe, the great Danish astronomer, made observations more precise than ever before. They enabled Johannes Kepler to deduce that the form of the earth's orbit around the sun is an ellipse. When the earth is closer to the sun, it moves faster, and when it is further away it moves slower. As a result, the length of a solar day varies through the year. Further complications arise from the tilt of the earth's axis, the obliquity of the orbit.

The unequal length of solar days is inconvenient. To simplify everyday life, we use mean time, with a fixed length of day equal to the average solar day. As a result, the sun is not due south at clock noon but sometimes running ahead and sometimes behind. The mathematical expression for this discrepancy is the 'Equation of Time'. The position of the sun at mean-time noon falls on a curve called an analemma. Mathematically, the analemma is a plot of the sun's altitude (angle above the horizon) versus its azimuth (angle from true north), and it has the form of a great celestial figure of eight.

Three adjustments have to be made to get mean time from sundial time. First, since Dun Laoghaire is just over six degrees west of Greenwich, 25 minutes has to be added. Next, a seasonal correction must be made. This is complicated to calculate, but help is at hand: it can be read from a graph of the Equation of Time, conveniently plotted on a bronze plaque. Finally, an extra hour must be added during Irish Summer Time.

The Dun Laoghaire Harbour Company is to be commended for installing the analemmatic sundial, 'a feature of artistic and scientific interest for all who use the amenities of the Harbour'. Local authorities elsewhere might follow this example. The sundial is a rich source of ideas for students, giving rise to many questions on geometry, trigonometry and astronomy, ranging from elementary problems to matters that have taxed the greatest minds.

The significance of the mathematical and astronomical theory involved in the Equation of Time is not confined to the design of sundials, but is important in many scientific and engineering contexts. It is used for the design of solar trackers and heliostats, vital for harnessing solar energy, which will one day be our main source of power.

FURTHER READING

The mathematical articles in the online encyclopaedia, www.wikipedia.org, are generally of very high quality. Many of these articles have been used during the writing of this book.

In addition, there is a great wealth of popular literature on mathematics. A few of the best books are listed here:

John H. Conway and Richard K. Guy, 1996: *The Book of Numbers*. Copernicus, New York. This is a fascinating compendium of the properties of numbers. It covers a remarkably wide range of topics in an understandable fashion.

John Derbyshire, 2006: *Unknown Quantity: A Real and Imaginary History of Algebra*. Atlantic Books. The entire historical development of algebra from ancient to modern times is told. The mathematical background is presented in separate chapters, making the book both self-contained and easy to read.

William Dunham, 1991: *Journey through Genius*. Penguin Books. Each chapter of this splendid book is devoted to a 'Great Theorem'. Mathematical details are interspersed with interesting historical context. I strongly recommend this book.

Marcus Du Sautoy, 2004: *The Music of the Primes*. Harper Perennial. This book traces the history of mathematics leading to the great Riemann Hypothesis. It is excellently written and is enjoyable and informative to read.

Richard Elwes, 2010: *Mathematics 1001*. Quercus Books. A comprehensive collection of very short articles on every aspect of mathematics. The book contains 1001 bite-sized and very readable essays, ranging from elementary mathematics to the most recent advances.

Ioan James, 2002: *Remarkable Mathematicians*. Mathematical Association of America. Biographies of 60 mathematicians from Leonhard Euler to John von Neumann.

David S. Richeson, 2012: *Euler's Gem*. Princeton University Press. A prize-winning book that traces the impact of Euler's polyhedral formula $V - E + F = 2$ on topology. The consequences and applications of this formula are many and fascinating.

Glen Van Brummelen, 2013: *Heavenly Mathematics*. Princeton University Press. Subtitled 'The Forgotten Art of Spherical Trigonometry', this book is a rich source on the history of astronomy and navigation. It breathes new life into a branch of mathematics that has great beauty but is no longer in fashion.

INDEX

acoustics 279-81

Alexandria 54, 55, 206, 248, 273

algebra 57, 145, 205-7, 209, 276-8, 289

Al-Khwarizmi 206

Ambassadors, The 291-3

analemmatic sundial 320-2

anamorphosis 291, 293

Andrews, George 39

anholonomy 186

Antikythera mechanism 121-3

Arabic science 205-7

Archimedes 43, 47-9, 54, 98, 116, 122, 224, 244, 301

Archimedes Palimpsest 48-9

Aristotle 92, 215-6, 248

astronomical perturbations 150-3, 155

atmospheric railway 101-3

Babylonian mathematics 67, 122, 137, 139-40, 250, 321

Banach, Stefan 246-7, 264

Barnsley fern 168-9

Barnsley, Michael 166, 168-9

Bayesian statistics 26, 61-3

Bayes' Rule 62

BBP formula for pi 97, 99-100

Beck, Harry 50, 52

Beethoven, Ludwig van 3, 4, 39

Bélanger's equation 135

bell curve 176, 266, 267

Beltrami, Eugenio 79

Bernoulli, Jakob 295

Bernoulli, Johann 295

Bernoulli's Principle 312

Bernstein polynomials 147-9

Bézier splines 146-9

Bézout's theorem 142-5

biology, mathematics in 2, 71, 191-3

Black–Scholes equation 265-7

Bolyai, János 79, 80, 115, 251

Boole, George 203, 289, 319

Box, George 26

Brahe, Tycho 13, 321

Brouwer's fixed-point theorem 264

bubbles, soap 224-8

Buffon's needle 285-7

Cantor, Georg 181-4, 189

carbon-60 118

cardinals, transfinite 183

Carroll, Lewis: see Dodgson, Charles Lutwidge

cartoon curves 199-201

Cassini, Dominique 218

Cauchy, Augustin-Louis 108

Cayley, Arthur 209, 289

Cayley table 238, 240

Champernowne number 86, 100

chaos game 166-9

Chester Beatty Library 207

circle of fifths 195-6

Clairault, Alexis Claude 219

Clay Mathematics Institute 30, 236

clothoid curves 157-9

cobweb plot 263

Collatz conjecture 94-5

complex numbers 143

computer proofs 23, 44-6

computer trading 265-7

conformal projection 271

Conway, John Horton 15, 140, 141, 179, 180

Courant, Richard 230

CT imagery 58-60

curvature 78-80, 114, 157-8, 178-9, 225-6, 228, 235, 270

cycloid curve 294-6

Dantzig, George 110-1

Descartes, René 31, 276, 278

dimensional analysis 202-4

Diophantus of Alexandria 140

DNA 60, 86, 223, 307, 308

Dodgson, Charles Lutwidge 45-6, 104, 106

dozenal system 67-8

drag force 216-7, 309-10, 311-13

Dürer, Albrecht 303

Dyson, Freeman 40

earth, hole through 104-6

earth, rotation of 185-7, 218, 294, 296

earth, shape of 53-5, 115, 218-20

eclipses 122-3, 155, 211

Einstein, Albert 37, 56, 104, 155-6, 178, 211, 214, 274, 275

epidemics, modelling 24-6

equals sign 56-7

equation of time 321, 322

Eratosthenes 53, 54

error-correcting code 297-9

Euclid 44, 45, 46, 51, 78, 82, 92, 113, 114, 132, 137, 250-2

Euclidean geometry 78, 113, 176, 248-52

Euclid's *Elements* 44, 116, 206, 248-52

Euler, Leonhard 50, 83, 116-8, 159, 243, 282-4,
Euler's polyhedron formula 116-8

factorial numbers 314-6
Fawcett, Philippa 89
Fermi, Enrico 202
Fibonacci numbers 170-2
fixed point theorems 262-4
forecast factory 317-9
forensic mathematics 2, 306-8
Foucault, Léon 185, 186
Foucault, pendulum 185-7
fractals 127-30, 131-3, 166-9
Fresnel, Augustin-Jean 159

Galbraith, Joseph 186, 187
Galileo, 2, 30, 104, 182, 203, 215, 216, 243, 256, 294
Galois, Evariste 276
Galois' theorem 241, 276
Gauss, Carl Friedrich 3, 4, 22, 78-80, 114, 115, 176-9, 243, 251
general relativity 37, 115, 155, 178, 209, 211-4, 274
GIMPS 30, 31
Global Positioning System 35-7, 220, 257
Goldbach's conjecture 83, 84, 94
golden angle 171, 253-5

golden number 171, 254
golf ball dynamics 310, 311-3
Google 1, 8-10, 129
Gosset, William Sealy 163-5
GPS: see Global Positioning System
graph theory 39, 50, 284
Graves, John T. 208, 209, 210
gravitational waves 155, 156
great circle equation 12, 13, 113
Grothendieck, Alexander 278
group theory 39, 72-4, 191, 237-41, 277
group velocity 75, 76

hailstone numbers 94-6
Hales, Thomas 22-3
Halpin, Robert 259
Hamilton, Richard 235, 236
Hamilton, William Rowan 75, 154, 208-9
Hamming, Richard 297
Hamming codes 298-9
Hardy, G. H. 38, 40, 89, 90
harmonic series 160-2, 174
harmonics (music) 194-8
Harriot, Thomas 21
Haughton, Samuel 186, 187
Hawking, Stephen 18, 56
Heaviside, Oliver 260-1
Hilbert, David 22, 147, 184, 274

Holbein, Hans 291
homeomorphism 51, 221, 236
Huygens, Christiaan 256
hydraulic jump 134, 135, 136

Ibec 170
infinity, degrees of 181-4
infinity, points at 143
Irish Times, The v, viii
iterated cosine function 262
iterated function system 168

Jordan curve theorem 16

Kelvin, Lord, 19, 76, 90, 222,
    259-60
Kelvin wake 75-7
Kepler, Johannes 13, 21, 22,
    23, 116, 151, 321
Klein 4-group 237-41
Klein, Felix 79, 239, 252, 274
knot theory 192, 221-3
Koch snowflake curve 127-8
Königsberg, bridges of 50,
    282-4
Kovalevskaya, Sofia 107-9,
    273

Lagrange, Joseph-Louis 22,
    275
Lambert, Johann 98
Leibniz, Gottfried Wilhelm von
    48, 295, 319

Le Verrier, Urbain 155
Liber Abbaci 170, 206
Lie groups 277
Lie, Sophus 277
Lindemann, Ferdinand von 98
linear algebra 8, 9
linear programming 110-1
Lobachevsky, N. I. 79, 80, 115,
    243, 251
London Underground map
    6, 50-2
Lucas–Lehmer primality test
    31
Lvov, School of 245-7

Mathematical Tripos 11, 18,
    88-90, 188, 288
Mathematics Genealogy
    Project 242-44
Maupertuis, Pierre Louis 219,
    220, 256
Maxwell, James Clerk 89
Melencolia I, 303-5
Mercator, Gerardus 271
Mercator projection 270-2
Mersenne, Marin 30
Mersenne primes 29-31
metric 113, 114, 137, 223
Mode-S EHS data 256-8
monster group 73, 74
morphogenesis 69-71
motorway design 157-9
musical scales 194-8

music, mathematics and 4, 194-8, 237, 241, 290

Napier, John 13, 319
Napier's rules 11-14, 65
Navier–Stokes equations 19, 313
navigation 32, 34, 35-7, 64, 113, 121, 220, 256
Neptune, discovery of 154, 155
networks 9, 25, 117, 191, 192, 193, 229-33
Neumann, John von 98, 111
Newell, Alan 171, 172
Newton, Isaac 2, 18, 48, 98, 132, 144, 150, 152, 212, 218, 243, 267, 275, 295
Newton's Laws 102, 105, 151-2, 154, 155, 212, 307
Nightingale, Florence 289
Noether, Emmy 72, 273-5, 277
non-Euclidean geometry 78-80, 113-5, 176-9
normal numbers 85-7, 99-100

O'Brien, Matthew 289
octonions 208-10
Ordnance Survey of Ireland 127, 130, 272

packing problems 21-3, 172
PageRank algorithm 8-10

Pappus of Alexandria 45, 46
parallel postulate 78, 79, 251
Paris, bridges of 282-4
Pascal, Blaise 295, 319
Pascal, Etienne 31
pendulum, simple 256
pendulum, Foucault 185-7
pendulum, to measure gravity 219
Penrose, Roger 98
Perelman, Grigori 235-6, 243
perspective in art 291-3, 304-5
perturbations 152-3, 155
phyllotaxis 171, 253
pi, computation of 47, 97-100, 286-7
Plateau's problem 224-8
Plato 92, 116, 248
Platonic solids 116, 250
Plimpton-322 tablet 139
pneumatic networks 319
Poincaré, Henri 183, 234, 235
Poincaré's conjecture 235-6, 243
Pólya, George 89-90
population growth 124-6
Prandtl, Ludwig 310
prime numbers 29-31, 54, 73, 82-4, 94
probability 8, 61-2, 85, 147, 161, 168, 176, 191, 243, 285, 286, 307, 308

projective geometry 91-3,
    292, 304
Pythagoras 92, 137, 195, 250
Pythagoras' theorem 65, 92
    195, 250
Pythagorean triples 137-40
Pythagorean tuning 195, 196

quaternions 208, 209

Radon, Johann 59
Ramanujan, Srinivasa 38-40,
    316
Ramanujan's lost notebook
    39, 316
Raphael's *School of Athens*
    92-3
Rayleigh, Lord 75
reaction–diffusion model 70
Recorde, Robert 56-7
relativity theory 37, 115, 179,
    211-4
reverberation equation 280
Reynolds, Osborne 309, 310
Reynolds' number 309, 310,
    312
Richardson, Lewis Fry 129,
    131, 317, 318
Riemann hypothesis 30
Riemann tensor 178
Riemannian geometry 137
Röntgen, Wilhelm 58

Sabine, Wallace Clement
    279-81
Scottish Café, Lvov 247
Seifert, Herbert 223
set theory 181-2, 188-9, 246
sexagesimal system 67
Shackleton, Ernest 32-4
Sierpinski gasket 166-8
sieve of Eratosthenes 54
simple harmonic motion 105
simplex method 111, 112
SIR model 25-6
sky-divers 216-7
slicing a pizza 268-9
slinky, falling 27-8
Smith, Henry John Stephen
    290
space, shape of 176-80
special relativity 37
spherical trigonometry 11, 12,
    33, 64-6, 114, 177
spiral curves 122, 158, 170,
    171, 263, 300-2
splines 146, 148-9, 159
sprouts, game of 15-17
Steiner, Jakob 229
Steiner's minimal tree problem
    229-33
Stirling's approximation 315-6
Stokes, George Gabriel 18-20,
    90
string theory 40, 74, 209-10,
    223

Student: see Gosset, William Sealy

Student's t-test 164-5

sudoku 5, 6

Sylvester, James Joseph 288-90

symmetry 4, 42, 72, 73, 105, 200, 210, 230, 273, 275, 277

systems biology 192-3

Taylor, Geoffrey Ingram 203, 204

telegraph equation 260

Theorema Egregium 80

thinking like a mathematician 6, 7

Thomson, William 19, 76, 90, 222, 259-60

tidal bores 134-6

tomography 58-60

topology 6, 16, 39, 50, 51, 115, 117, 192, 221, 223, 231, 234, 242, 277, 284, 324

trigonometry 11, 64, 306, 307, 322

trilateration 36

Trinity atomic test 203

tunnel through the earth 104-6

Turing, Alan 69, 70

ungula, volume of 42, 43

Venn diagram 188-90, 298, 299

Venn, John 188

Verhulst, Pierre François 125

watermelon puzzle 119-20

weather prediction 2, 19, 256, 317, 318

weather records 160-2

Weierstrass, Karl 108, 147

white holes 135

Wolfram Alpha 199

world population 124-6

Wrangler, Senior 18, 88-90

Wren, Christopher 295, 301

Wren Library 39

X-ray imaging 58-60

Young–Laplace equation 225

Zariski, Oskar 278

Zhang, Yitang 83